# technocreep

THE SURRENDER
OF PRIVACY AND
THE CAPITALIZATION
OF INTIMACY

THOMAS P. KEENAN

# techno creep

GREYSTONE BOOKS
*Vancouver/Berkeley*

*To my parents, Ruth and Joseph, for allowing me to have white rats and cancer viruses in our basement at the age of fourteen. To the love of my life, Keri, who inspires me every day with her amazing Australian wit and wisdom; and to my wonderful son, Jordan, who bounces ideas around with me like a pro and is navigating his own amazing path in life.*

Published by arrangement with OR Books LLC, New York
Copyright © 2014 by Thomas P. Keenan

15 16 17 18 19 8 7 6 5 4

All rights reserved. No part of this book may be reproduced, stored in a retrieval system or transmitted, in any form or by any means, without the prior written consent of the publisher or a license from The Canadian Copyright Licensing Agency (Access Copyright). For a copyright license, visit www.accesscopyright.ca or call toll free to 1-800-893-5777.

Greystone Books Ltd.
www.greystonebooks.com

Cataloguing data available from Library and Archives Canada
ISBN 978-1-77164-122-7 (pbk.)
ISBN 978-1-77164-123-4 (epub)

Front cover and text design by Bathcat Ltd.
Typesetting by Lapiz Digital
Printed and bound in Canada by Friesens
Distributed in the U.S. by Publishers Group West

We gratefully acknowledge the financial support of the Canada Council for the Arts, the British Columbia Arts Council, the Province of British Columbia through the Book Publishing Tax Credit, and the Government of Canada through the Canada Book Fund for our publishing activities.

Greystone Books is committed to reducing the consumption of old-growth forests in the books it publishes. This book is one step towards that goal.

# Contents

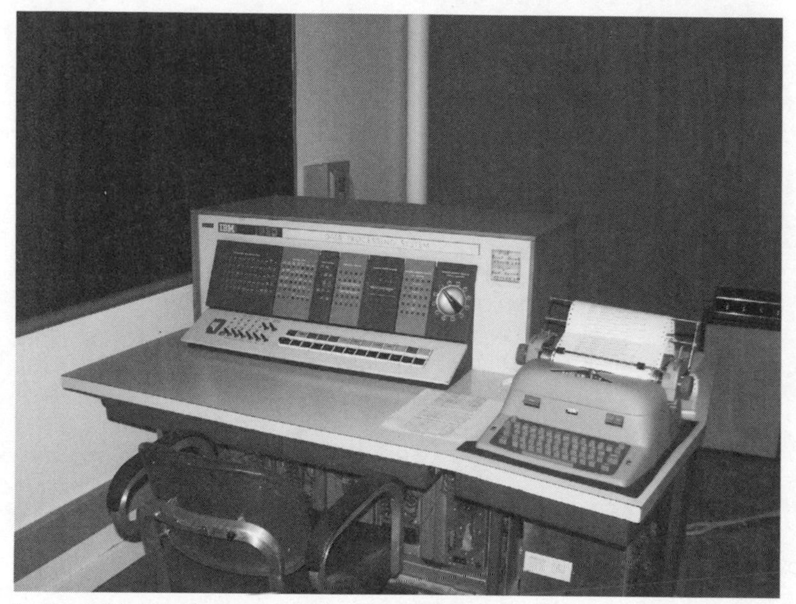


Figure 1. IBM 1620 computer like the one at Bronx Science. Erik Pitti, via Flickr/Creative Commons Attribution License.

# Preface

I wrote my first computer program in 1965, while I was a student at the legendary Bronx High School of Science. Tech pioneers such as Marvin Minsky, Robert Moog, and Martin Hellman once walked its halls, which sometimes reeked of chloroform. In those days, students were actually allowed to perform surgery on small mammals. I remember coming in early one morning to help my friend Mark remove the spleens from several hapless white rats.

Bronx Science had a computer, then a rarity in all but the largest businesses and almost unheard of in a school. It was a cranky card-munching monster that we regarded with a combination of veneration and lust. In retrospect, jockeying for time on a computer seems like a bizarre hobby for a group of normal teenagers. However, the students at Bronx Science were anything but normal. Eight alumni have gone on to win the Nobel Prize. Two of the Nobel laureates in physics, Russell Hulse and Hugh David Politzer, were in my graduating class. I was in good company.

Access to the school's computer was strictly controlled. Only seniors were allowed near the hallowed IBM 1620 console. We juniors were forced to sit for hours in front of whirring calculators, doing endless numerical analysis calculations and writing down the answers. Our teachers hoped that this would help us appreciate the magical day when we finally got to put our little deck of carefully punched cards into the whirring IBM 1622 Card Reader/Punch. The computer would then do the calculations we had slaved over for the last term in mere minutes, or even seconds.

Driven to get my hands on a computer sooner, I discovered a special youth training program at New York University. If we were willing to give up our evenings and weekends, the folks there would teach us all the computer programming we could possibly absorb. We were even allowed to leave our card decks, secured with rubber bands, for the computer operators to run when they had nothing better to do.

The door of the building that housed NYU's computer had a sign that said "United States Atomic Energy Commission." There was a curtain to shield the IBM 7094 from prying eyes when it did secret work. The Soviets had launched Sputnik in 1957, leading to fears that they would dominate the world from space. The 1962 Cuban missile crisis had us doing air raid drills in school. Clearly computers were going to play a role in saving America, and we were being trained to play a part in that drama.

Under the leadership of professors including Max Goldstein and Jacob T. ("Jack") Schwartz, and coached by a kind-hearted and energetic NYU researcher named Henry Mullish, there were no limits to what we could accomplish. We created whole new computer languages, fixed bugs in existing ones, and wrote emulators for computers that were still on the drawing boards. I wound up programming everything from the statistics for numerous PhD dissertations to particle physics calculations to some of the structural engineering data for the original World Trade Center. I earned my keep on that last project by catching a glitch that might have caused the two 110-story towers to collide in high wind conditions.

Back then, computers spat out their results in 132 column–wide format on oversized continuous sheets of paper. When I pulled out a printout while riding the Bx40 cross-town bus in the Bronx, I always got strange looks from the other passengers. I was proud that I had something special and almost magical in my hands. I now realize they probably thought I was a very creepy kid.

Years of being a computer programmer, a computer science professor, and a technology journalist have helped me realize that almost every new technology can be misused and often become deeply disturbing. In 1984, I had the great fortune to co-write and host a CBC IDEAS series called *Crimes of the Future*, in collaboration with

Dr. Duncan Chappell, then head of the Criminology Department of Simon Fraser University; and Dave Redel, a very talented CBC Radio producer.

Those programs marked the first time many people heard about identity theft, except perhaps in the context of someone going to a graveyard to copy the name and birth date of a deceased infant. We talked about crimes that have now become real, such as trafficking in human body parts, and others that are only now surfacing, like "wireheading"—the direct stimulation of the pleasure centers of the brain. Back then, we had to use the work of science fiction writers like Larry Niven and Spider Robinson to introduce this intriguing practice. Today there are detailed, instructions for brain self-stimulation on the Internet.[1]

For almost five decades now, I have been watching as everybody else gets on the tech bandwagon, sometimes adroitly, sometimes clumsily, and often without fully understanding the implications of what they are doing.

This work has taken me to conferences like DEF CON; Black Hat; Computers, Freedom, and Privacy; and led to unexpected adventures such as being allowed to scrub in on a liver transplant operation. I have had the privilege of talking to thousands of academics, visionaries, and technology creators, and come away with the strong sense that we need to raise the bar in our thinking about technocreepiness, and sooner rather than later.

Over half the people on that Bronx bus today would now be glued to their smartphones, connecting with friends, checking sports scores, or enjoying celebrity gossip. Would any of them be thinking about the next chapter in technology, and how it will change their lives?

This is a book about the creepy pioneers of technology: what they are doing and why we need to know about it. In the pages that follow we will go on a journey into some of the disturbing ways our lives are unfolding, often behind the scenes and without our knowledge or permission. Not all of these incursions are necessarily bad. The benefits of knowing you are prone to a certain disease, for example, *might* outweigh the risks of having genetic tests on your medical record. And what seems creepy today may be the accepted norm in the future.

Still, there is this nagging feeling that decisions we make today may come back to haunt us in the future, in ways that are hard to envision. Yet that is precisely what we should be doing.

# Introduction

Modern technology is not what it seems. Or rather it is much more than it seems. Digital wheels that most people do not even know exist are turning in the background. Increasingly, we are getting an uneasy feeling about this ... a sense that things are not quite what they seem.

So much is happening that is out of our view and beyond our control. Like a network of mushroom spores sending out subterranean tendrils to silently exchange genetic material, our technological systems are increasingly passing information back and forth without bothering to tell us. They are parsing and analyzing it to squeeze out the deep meaning of what we say and do, sometimes before we are even aware of our own intentions.

Technocreep is quietly but relentlessly invading our daily lives:

- You use your smartphone to take a photo, and it auto-uploads it to Facebook. Without your knowledge, metadata such as the type of camera you use and the precise location where you took the photo is also being uploaded. Facebook may remove that information before making your photos public, but the company certainly has access to all of that metadata for its own purposes. Facebook now has the world's largest known database of personal information and photos, many of them conveniently labeled with your real name. How deep is the analysis of your words and photos by Internet giants like Facebook and Google? According to an article in *Wired*, the computers at Facebook can use artificial intelligence to tease out the emotions in your ramblings and figure out when you are being sarcastic.[2] That same article says that Google can distinguish the facial features of a cat from a human.

- You decide to check your email, which, like most people, you are now getting for free from a provider like Google, Yahoo, or Microsoft. Hmm, no new mail has been delivered for the past few hours. Do their servers get backed up like the post office at the holidays? Or is something more sinister going on? Is your mail being siphoned off for some sort of deep-level human or computer analysis? Given the revelations of Edward Snowden and others, your concern might be justified. What should certainly disturb you is the fact that you have no way to really know how your web-based email is processed, and virtually no tools to investigate it; while others seem to have many tools to investigate you.
- Before heading to bed, you peruse the electronic catalog of an upcoming estate sale, lingering on the image of a nice chandelier. An advertisement for the website www.chandelier.com pops up on your screen. How did they know what you were thinking about? Perhaps it was because the sale image was saved as chandelier.jpg. But perhaps not. Image recognition technology is progressing at an amazing pace. Again, the wheels are turning in the background in a creepy fashion.
- Late at night, you hear the hard drive whirring on your computer. The monitor is flickering even though nobody is using it. Perhaps it makes a few of those strange "bonging" sounds that signify someone is sending you a message. You look. But there is nobody there, and the computer, as if sensing your presence, has ceased its frantic activity. All is calm. But you are not sure if it was Microsoft doing software patches, a hacker trying to steal your information, or … something else.
- Bars in several cities have installed cameras that silently watch their clientele and make inferences about them from their physical characteristics. Armed with the free smartphone app SceneTap, prospective patrons can check out how full the place is ("chillin" to "hoppin"), as well as the average age and percentage of males and females there.

- "Suggestion algorithms" are popping up on shopping and social networking sites. It is no surprise that Amazon is trying to sell me the last few things I price-checked there (and bought elsewhere). But when it suggested "adult size disposable diapers" as a good purchase for people buying certain video games, did it "know too much"?
- Google's Regina Dugan has suggested with a straight face that you may soon swallow a password pill or sport a digital tattoo to log on to computer systems.[3] Both ideas are technologically feasible right now, but should our employer be allowed to brand us or make us take a pill?
- Next generation wearable computers such as Google Glass may start regularly tracking where you are looking.[4] That information will then be sold to advertisers and others who are seeking a window into your mind. Based on where your gaze lingers, they will suggest things you want to buy even before you know you want them.
- Your phone may listen for audio cues about where you are. Is that a football stadium announcer it hears? Perhaps you would like a discount coupon for the team's store, or, if you are slurring your words, a connection to a "drive me home in my car" service.

Perhaps you will seek refuge from all this invasive technology in some nostalgic low-tech activity like attending a rock concert. We will still have those in the future, as people clamor for the "live experience" in a world with unlimited digital media access. But the fiftieth anniversary Woodstock concert will probably be a lot different from the original one in 1969.

At Woodstock 2019, you may be swatting away disposable flying robot cameras that people have launched to catch a better view of the performers. Folks all around you will be pushing the "find my friends" button, revealing their exact location at the concert venue. Unlike today's friend finder apps, the 2019 version may also tap into their brain waves and body chemistry to decide if they are interested in joining you for some food, or perhaps something else. Yes, babies will be conceived at Woodstock 2019 just as they were in 1969.

If you have broken some minor rule like forgetting to renew your vehicle's license, you will come back at the end of the show to see your "smart license plate" displaying "EXPIRED" where the number should be. Should you decide to buy a souvenir T-shirt, your every move will be tracked by cameras that make uncannily accurate estimates of your age and gender, and then predict your buying habits.[5] They may even recognize you by your face or change the prices based on the color of your credit card, which it will be broadcasting to the world.

Should you seek medical help at Woodstock 2019, you will almost certainly receive tests and treatments tailored to your unique genetic makeup. Dr. Leroy Hood, the biologist who pioneered automated DNA sequencing, describes what you should expect at the Woodstock 2019 first aid tent. "We'll be able to prick your finger," he says, "and take a droplet of blood and make ten thousand measurements that tell us about a gazillion different things that may speak to why you have apparent cardiac pain. There'll be very powerful imaging devices that can probably be done at Woodstock that could look at the brain or look at the heart."[6]

Hood also believes that this type of personalized medicine will come down in cost very rapidly. "My own prediction," he says, "is with third generation DNA sequencing ... the genome will cost $100 and we'll be able to do it in fifteen minutes." That price drop, from the current level of thousands of dollars, should put your mind at ease. After all, there will be no way to skip out on your bill at this medical facility. They will have your DNA sample.

Unlike in 1969, when misadventures with psychedelic chemicals accounted for many visits to the medical tents, the Woodstock 2019 medics may be treating the after-effects of electronic stimulation of the pleasure centers of the brain. And just as crowdfunding sites are starting to determine which products are manufactured by online consensus, Woodstock's organizers may be able to tap into the "hive mind" of concert-goers to keep the show moving at just the right pace.

If you cannot make it to the live event, there will be ultra-high-resolution videos to enjoy, streamed directly into your retina or perhaps even your brain. Organizers will probably use big data analytics and artificial intelligence to choose the lineup of musicians. That is how Hollywood already decides which television shows we are most likely to watch.[7]

This book is about the unseen ways in which technology is already changing our lives. We will visit the hotel suites at the DEF CON and Black Hat conferences, where hackers attack circuit boards and tweak software late into the night. We will go into the online nooks and crannies where digital exploits are secretly shared. We will even examine a kids' toy that frightened the National Security Agency so much that it was banned from their building. You will learn why you might want to avoid certain kinds of medical testing and why your online presence will definitely outlive you, unless the "transhumanists" are correct and the first immortals are already living among us.

Many people believe that these disturbing technologies are confined to the Internet and that if they are careful, or even avoid online activity altogether, they will be safe. But the technologies that will truly change our lives will be in our cars, our streetlights, our hospitals, and even inside our brains and bodies. Our favorite watering holes, even our pets and our children, are being infested with technocreepiness.

The general public learns about creepy technologies episodically. A whistle-blower unveils whole areas of government or corporate snooping. A probing journalist figures out how big data can be used to link together nuggets of your life to create a chillingly accurate portrait of you. A scientist works backwards from a DNA sample to infer the most likely surname of the person it came from.[8] Perhaps you receive a particularly astute yet unsettling suggestion for a new contact on Facebook or LinkedIn, and wonder how they did that.

Every new technology eventually attracts calls to restrict or regulate it, and often to find a way to turn it into a revenue stream and make it taxable. But the flow is becoming too fast, too diverse, and too imaginative for lawmakers to keep up.

We will always have new innovations, and people will find ways to misuse some, and to combine them in unanticipated ways. Some ideas will pass into oblivion like the pernicious RottenNeighbor.com. For a while, this website let you anonymously badmouth folks whose dogs allegedly pooped on your lawn or falsely label the guy down the street as a sex offender. Companies were not exactly eager to advertise on RottenNeighbor.com, and it now sits, dormant, though still registered and presumably ready to rise again if a viable business model suddenly appears. Gossip sites like TheDirty.com continue to provide an outlet for this kind of vicious personal attack, often tied to your geographic location.

Creepiness is an elusive concept that taps into our primal fears and assumptions about the way things are and should be. Sigmund Freud pondered it, suggesting that creepy things, with part of their true nature hidden, remind us of our own deepest secrets and guilt over repressed impulses. Edgar Allan Poe achieved heights of creepiness in his short stories by playing on our fears of being buried alive or catching the plague.

Filmmakers often seek the fine balance that will send the audience away wondering darkly about a character's true nature and motivation. Culture blogger Sarah Dobbs explains that movies can be scary with loud noises and sudden moves, but that truly unsettling cinema is a lot harder to achieve.

"To be one of the good horror movies," she writes, "a film needs to establish a certain atmosphere; it needs to draw you in and make you care. It needs to give you something to think about when you're trying to drop off to sleep at night; to make you wonder whether that creaking noise down the hallway was just the house settling, or something lurking in the shadows. Creepy stays with you. It gives you goosebumps."[9]

Clowns, dolls, ghost stories, and even the words from the mouths of our young children can fascinate us in unsettling ways. The closer something is to our hearts and our highest values, the more disturbing it can be.

A famous discussion on the social news site reddit asked, "What's the creepiest thing your young child has ever said to you?"

Here are two of the more disturbing examples:

*I was tucking in my two-year-old. He said, "Goodbye, Dad."*
*I said, "No, we say goodnight."*
*He said, "I know. But this time it's goodbye."*
*Had to check on him a few times to make sure he was still there.*
*/u/UnfortunateBirthMark*

And

*When I was about three we had a cat that had stillborn kittens.*
*I asked my father if we could make crosses for them, which he did.*
*As he was making them I asked: "Aren't those too small?"*
*Dad: "What do you mean?"*
*Me: "Aren't we going to nail them to them?"*
*Dad (after several moments' silence): "We're not going to do that."*
*Me: "Oh."*
*/u/Tom_Zarek*[10]

Cute? Innocent? Perhaps. But the creepiness stayed with these people long enough for them to share it on reddit. The fact that this thread is now up to around 15,000 comments speaks to our fascination with the *unheimlich,* a German word that literally means "the opposite of what is familiar or home-like." By thinking about what unsettles us the most, we are able to confront and understand our greatest fears, and try to make rational decisions as citizens, software designers, creators, parents, and consumers.

So, what is creepy? We know it when we see it. The hairs stand up on our neck or we ask, "How do they know that?" Creepy things make us question our assumptions, and lie awake, wondering "what if?"

My study of hundreds of technologies that are creepy in various degrees has revealed some common factors that make people uneasy. At the end of this book, we will explore these "dimensions of techno-creepiness" in the hope that we can avoid them in the future. We will even do a bit of "creep-proofing" to, as much as is humanly possible, protect ourselves from the worst ravages of invasive technologies. But for now, let us let those hairs do their job. We will begin our journey into technocreepiness at a rather unlikely place—the New York Public Library.

# Intelligence Creep

There are twenty-nine steps from the corner of 41st Street and Fifth Avenue to the front entrance of the New York Public Library. I know this because, in the mid-1970s, I lugged a radio station's tape recorder up every one of them. I might as well have been carrying a television set. Back then, "portable" meant that something had a handle. I was there to interview the keeper of an amazing new technology—the Kurzweil Reading Machine.

Billed as an aid to the blind, this bulky contraption was the world's first functional text-to-speech synthesizer. Walter Cronkite used the machine to sign off on his January 13, 1976 newscast. I typed up a piece of paper with "For CBC Radio, this is Tom Keenan in New York." The machine rattled this off for me in the same mechanical monotone that we now associate with the hacktivist group Anonymous.

I then asked the librarian, "What kind of things do people bring in to read on it?"

"Mostly pornography," he replied.

I thought I had heard him wrong. He explained that "if somebody wants to hear a textbook on American History or something, there are plenty of volunteers who will read that. We're seeing books like *Lady Chatterley's Lover*, *The Story of O*, that sort of thing."

A lot has changed since my first encounter with the Kurzweil Reading Machine. Instead of lugging a bulky tape recorder, I can now push a button on my smartphone and safely store my interview in the cloud. If I want to know how many steps I will face at the New York Public Library, I can simply count them on Google Street View. Anyone with Internet access can find all the pornography they could possibly want, and have it read to them in whatever exotic voice they desire.

The abundance and variety of Internet pornography illustrates a concept that Cullen Jennings, one of my former students and now a Cisco Fellow, expressed very well. "No matter how liberal or broad-minded you are," he once said, "I guarantee I can find something on the Internet that will instantly offend you." It is a small leap from "something that will offend you" to "something that will creep you out."

Even though I have seen a lot of bizarre things since the 1970s, the image of tumescent guys hooked up to the Kurzweil Reading Machine at the public library has stuck with me, along with the gadget's monotone voice.

A good friend and I used to split the generous "lead fees" which a certain national tabloid paid for ideas that turned into stories. Driven by empty wallets, and armed with a bottle of Jack Daniel's, we could spin off quite a few plausible if sensational ideas in an evening. Of course, a tabloid tale is not a publishable story without a reputable expert to support it. This paper had a helpful list of "trained seals" who, for a fee, would happily confirm UFO sightings and authenticate photos of fictional monsters. Yet they had a big gap—they needed a bright young computer science professor, which was precisely my line of work.

Soon I came to be quoted on fantasy technology stories like "an amazing implanted chip will someday measure your caloric intake and release an appetite-suppressing hormone." That feature attracted bags of mail from people desperately seeking to help me test this rather wacky idea. I sent the letters back suggesting that they look into some diet and exercise plans.

Perhaps I should have told them to wait. An article in the 2009 issue of *MIT Technology Review* describes a "small, stick-on monitor no bigger than a large Band-Aid" that can accurately monitor your caloric intake.[11] Some smart scientist will undoubtedly invent the appetite suppression technique and make our fictitious dieter's dream patch a reality.

In that weekly tabloid, I also mused that "someday computers will speak to you in the voice of your choosing. It might be Marilyn Monroe's or Clark Gable's or the voice of your long-dead mother." Since then, science has shown that the sound of your mother's voice indeed does have a profound physiological effect on you.

In studies of mother/daughter dyads, Leslie Seltzer found that hearing Mom's voice raised girls' oxytocin levels, calming them down.[12] Email and SMS messaging did not have the same effect. Some have even speculated that an artificially intelligent program that sounded like your mother, used her favorite expressions, and had an intimate knowledge of your life story, could be a powerful way to calm, interrogate, or even control you.

Back in 1966, Massachusetts Institute of Technology professor Joseph Weizenbaum unleashed ELIZA onto the world, hoping to illustrate the fundamental differences between human and computer brains. A fairly simple "chatterbot," ELIZA could mimic the conversation style of a Rogerian psychotherapist, faking the answers when it did not understand what you just said.

Weizenbaum observed that a session with a psychotherapist was one of the few environments where it was perfectly acceptable to turn a question back on the speaker, or provide a canned response like "So, how does that make you feel?"

Although Weizenbaum intended ELIZA to be a parody with a point, many people enjoyed pouring out their personal problems to the program. Some even said they preferred it to a human therapist. When Weizenbaum told his secretary, who had been using the software, that he had access to the logs of all her conversations with ELIZA, she was outraged at this invasion of her privacy.[13]

ELIZA got herself into some humorous situations. In the age before massive hard disks, computer data was usually stored on 2,400-foot reels of magnetic tape which had to be manually mounted by an operator. The tape mount request came up as a flashing message on the operator's console.

However, some of MIT's computers did not have a human operator available. Pranksters configured one of them to use ELIZA to respond to operator requests. "Can you please mount a tape for me?" was met with "Is it because of your mother that you want me to mount a tape for you?"[14] Profanity usually ensued from the hapless human.

The idea of programs that can mimic human conversation has spawned some creepy chatterbots at major companies. In 2006, The Coca-Cola Bottling Company of Atlanta, GA, "hired" a virtual representative named Hank. Hank was intended to, among other things, deflect phone calls to the company's main switchboard from students doing reports on the history of Coca-Cola. He straddled the line between being a computer program and a person quite adroitly for his era.

I had some great talks with Hank. We discussed his personal life ("virtual representatives are not allowed to have relationships") and he vehemently denied that Coca-Cola has ever contained cocaine.

Other people also tormented Hank with questions like[15]:

Q. *Do you like snorting coke?*
A. *Of course! I like all of the products of The Coca-Cola Company.*

Q. *Are you a Mormon?*
A. *We are not in a position to comment on the ownership of stock. The holdings of our share owners are confidential. In accordance with our current governmental regulations, however, we include in our annual proxy statement a list of those share owners who are beneficial owners of five percent or more of our company stock, and the Mormon Church is not among them.*[16]

History does not record when Hank was downsized from a full-fledged virtual representative to a "frequently asked questions" webpage.[17] That seems to be all that is left of him today, though he is still wearing his trademark telemarketer headset.

If you could talk with Hank about such personal matters, why not unburden your venal and even mortal sins to a piece of technology? Online confession does seem to have some kind of creepy visceral appeal to people. In 1984, an art project called "Interactive Catholic Confessional"—based loosely on ELIZA—was put on display. "Visitors entered a confessional space, knelt before the computer and went through the process of Catholic confession," according to a posting at the University of Nevada at Reno. "The computer gave each user an appropriate penance for their sins."[18]

The concept of computerized confession is still with us. In 2013, a Jewish congregation in Florida urged its members to atone for their sins through anonymous, but very public, tweets which would scroll by on a screen during the Yom Kippur services.[19]

On a more secular level, reddit has a whole subreddit called r/confession, and entire websites like www.truuconfessions.com ("your anonymous best friend") thrive on this compulsion to share guilty secrets. Here, you can learn who is lusting after his cousin's wife and who "kicked a child (who probably deserved it)." These posts make fascinating reading, but of course their real purpose is catharsis for those who write them. An amazing number of people seem to spend a lot of time poring over these stories, "upvoting" and "downvoting" them, and adding their own commentary. In one sense it is a new way of communicating with a higher power, even if this higher power is only a transient, anonymous online community.

The line between machine and human thinking is definitely blurring, as is well illustrated by the triumph of IBM's Watson over the best human Jeopardy players. Virtual assistants like Apple's Siri and Microsoft's Cortana are mining our smartphones and emails to do some of our thinking for us.[20] We can feel the hot breath of our technology pushing us relentlessly towards that much-touted "singularity"—the day when our creations will be smarter than us in ways that really count.[21]

In the mid-1800s, Augusta Ada King, Countess of Lovelace, studied the work of Charles Babbage, who designed a precursor to modern computers. Because she wrote down the steps to compute the Bernoulli numbers on Babbage's never-constructed Analytical Engine, Lovelace is often called the first computer programmer.

She is also known for her famous "Objection" to the idea that a machine can possess creativity. "The Analytical Engine has no pretensions to *originate* anything," she wrote in 1842. "It can do *whatever we know how to order it* to perform"[22] (her italics).

If she were alive today, Lovelace might have trouble maintaining her position as Watson trounced her in her choice of intellectual games. But we do know that, putting aside quantum computers, neural networks, and other specialized technologies, mainstream computers still sequentially execute instructions that were designed by their human masters. If Lovelace's Objection is as true as ever, why do technologies do things that amaze us and give us creepy spinal shivers?

One explanation of this apparent paradox is that many computer programs have already surpassed the comprehension of any *one* human mind. This was actually true of the operating system, OS/360, made for IBM's mainframe computers in the 1960s and 1970s. It had so many modules and complexities that it took a team of systems programmers to build it, and nobody purported to know every inch of it. Mix in the creative input of today's very bright designers and programmers, and you get a continuous stream of technologies that amaze, delight, confound, and, increasingly, disturb or even frighten us.

Sometimes we have trouble detecting the boundary between machine and human intelligence. Most people recognize that the "Recommended" suggestion list from Amazon.com comes from a robot. But what about the earnest email appeal from a friend who claims he is stranded abroad without funds. The sender seems to know intimate details about your mutual relationship. It is probably a nasty robo-scam, but how can you be sure?

In his signature essay on the subject, Sigmund Freud tackled the psychological aspects of our discomfort with things that may or may not be human. Using the example of the doll in the first act of Offenbach's opera, *Tales of Hoffman*, he acknowledges that "doubts whether an apparently animate being is really alive" can invoke The Uncanny. Freud goes on to suggest that what we truly dread here is an Oedipus-style gouging out of our eyes, or, this being Freud, a symbolic castration.[23]

Japanese robotics professor Masahiro Mori coined the term "uncanny valley effect" to explain why we become uneasy when non-human things exhibit human-like behavior.[24] Perhaps nothing embodies the spirit of the uncanny valley better than BINA48.

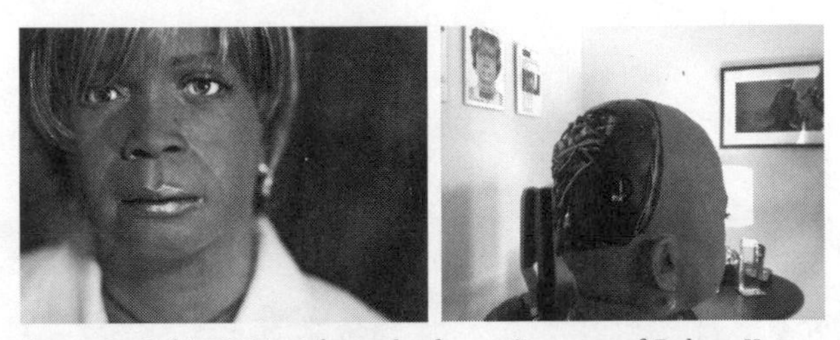

Figure 2. (left) BINA48 from the front. Courtesy of Robert Koier.
Figure 3. (right) BINA48 from the rear. Courtesy of Terasem Movement Foundation.

Martine Rothblatt, a serial entrepreneur, lawyer, and researcher, has created an extremely lifelike humanoid robot in collaboration with robotics engineer Dave Hanson. In addition to having a convincing and expressive face made of a polymer called "Frubber," BINA48 has an uncanny ability to display human mannerisms. *New York Times* reporter Amy Harmon, sent to Vermont to interview BINA48, reports a profound moment as BINA48 looked her in the eyes and said "Amy!"

"Maybe it was the brightening of the sun through the skylight enabling her to finally match up my image with the pictures of me in her database," Harmon writes. "Or were we finally bonding?" The spell was broken by BINA48's jarring next remark, which was to change the subject: "You can ask me to tell you a story or read you a novel."[25]

BINA48 has cameras in her eyes and is equipped with face finding and facial recognition software. As their cost plummets to virtually zero, digital cameras are turning up almost everywhere. They now seem to be present at the best, worst, and creepiest moments of our lives.

# Camera Creep

On April 19, 2013, law enforcement agents used a thermal imaging camera, combined with a tip from a citizen, to locate Dzhokhar Tsarnaev in the aftermath of the Boston Marathon bombings. The image of a human form crouched under a tarp sped around the globe. The imaging technology was praised for leading to the result almost everyone was hoping for: the live capture of a desperate fugitive.

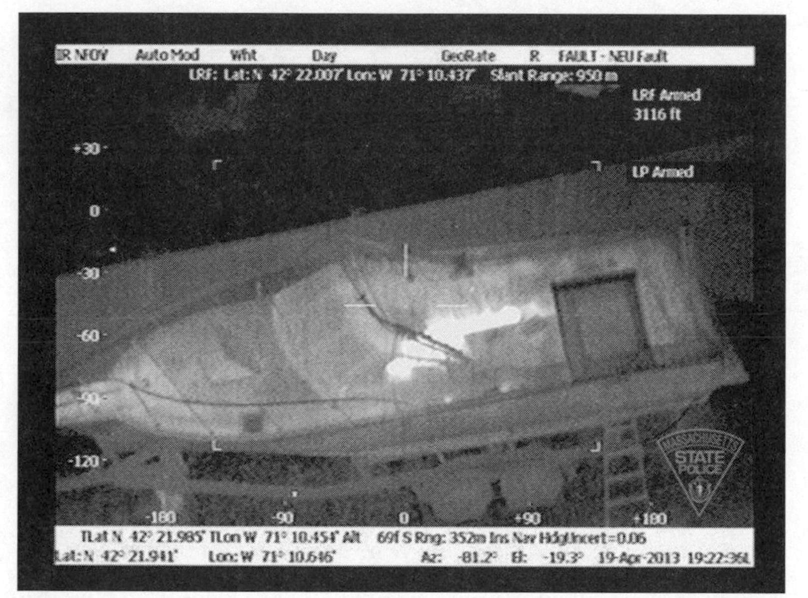


Figure 4. Fugitive Dzhokhar Tsarnaev, hiding under a tarp. Courtesy of Massachusetts State Police Air Wing.

Thermal imaging cameras are not new. They have been used for years by firemen (who look for cool spots since burning walls are much hotter than trapped humans) and by house inspectors probing for heat-wasting leaks.

They also play a role in tracking down marijuana grow-ops and finding people and objects hidden in walls and vehicles. Yet, suddenly thermal imaging was front page news, seeming to give law enforcement superpowers. From their helicopter, the Massachusetts State Police found a needle in a haystack, using what seemed like a kind of X-ray vision.

Regular cameras also played a role in this investigation, as agents pored over masses of amateur cell phone video and surveillance camera footage. The cameras that yielded the best pictures were the ones mounted on the Lord & Taylor store and the Forum Restaurant at 755 Boylston Street in downtown Boston.[26]

I walked that very stretch of Boylston Street a few months earlier. While those cameras were not hidden, they certainly would never catch your attention. Yet they provided vital evidence. After sifting through all the images, authorities published photos of suspects they dubbed "Black Hat" and "White Hat," asking for the public's help.

The "go public" strategy worked, and Dzhokhar Tsarnaev was soon apprehended alive. Despite this outcome, some academic researchers called the Boston Marathon bombing case "a missed opportunity for automated facial recognition to assist law enforcement in identifying suspects."[27] Joshua C. Klontz and Anil K. Jain of Michigan State University did an after-the-fact simulation using the Boston suspect photos and a database of one million mugshots released under Florida's "sunshine" law. They stirred in photos of the Tsarnaev brothers taken on various occasions such as after a boxing match in 2009. Using commercial facial recognition software, they had some success in matching them, including a "rank one" matchup between a bombing scene photo released by the FBI and a high school graduation photo of Tsarnaev.

The researchers acknowledge that neither of the commercial facial recognition systems they tested is ready for routine deployment in law enforcement applications, largely because of issues with different poses, resolution, and factors as simple as wearing sunglasses.

However, you can be sure that work on improving facial recognition for law enforcement is moving full speed ahead.

Soon after the Boston Marathon bombings, Joseph Schuldhaus, vice president of information technology for Triple Five Group, which runs the sprawling West Edmonton Mall as well as Minnesota's Mall of America, suggested that video analysis is going to become even more important in fighting crime. "I think we're going to see the further miniaturization of algorithms at the edge of the device," he told IT World Canada editor-at-large Shane Schick, "and what I mean by that is when the video comes into the camera these algorithms are going to help law enforcement better process that information, much like when you use Shazam to identify a song."[28]

Schuldhaus clearly believes that the public safety and security advantages of surveillance cameras outweigh the risks they pose to privacy. Others are not so sure; but this has not stopped cameras, both public and private, from proliferating around the world.

According to a report in *Forbes*, "In the United States, it is estimated that there are 30 million surveillance cameras, which create more than four billion hours of footage every week."[29] They are also sprouting a lot of intelligence and new functionality. Their images are processed, in real time, to highlight suspicious packages at airports, to discover people who go where they are not supposed to be, and, even, as proposed by some Japanese researchers, to catch kids smoking in the schoolyard.[30]

For many years, a conference called Computers, Freedom, and Privacy featured a post-conference tour of the host city's surveillance cameras. I vividly remember going on the tour of San Francisco in 2004. We stopped after we found about a hundred cameras peering down at unsuspecting people in Union Square and other public venues.

Dedicated volunteers from the New York Civil Liberties Union walked around Manhattan in 1998 noting camera locations, producing what they called "a comprehensive map of all 2,397 surveillance cameras in Manhattan." When they re-did the same study in 2005, they "found 4,176 cameras below Fourteenth Street, more than five times the 769 cameras counted in that area in 1998."[31]

They are fighting an uphill and ultimately hopeless battle. Modern surveillance cameras can be tiny, totally wireless, solar powered, and cost a few dollars. Good luck spotting one of those pointing out of a window or hiding in the pore of a ceiling tile.

It is hard to deny that the presence of video cameras in public places has deterred some criminals and solved or prevented certain crimes. In one case, a bandit robbed a local wholesale club, and was caught on the surveillance camera. A simple scan through the membership records turned up a match for his photo, yielding his name and home address.

The heavy camera coverage of New York's Times Square is credited with helping police thwart the plot to detonate a bomb there in 2010. However, just as in the Boston Marathon bombings, a tip from a citizen also played a major role. Once, while watching a camera pointed at Times Square, I saw an entire drug deal transpire in plain sight.[32] I managed to capture several screen shots and use them as an example of people who obviously did not realize they were being watched. Or who did not care.

Do cameras really earn their keep as crime fighters? The best data on this comes from the United Kingdom, which has had extensive camera coverage for over a decade. The results are not as encouraging as camera advocates would like us to believe. A 2009 Scotland Yard report estimated that only one crime was solved per year per thousand cameras.[33] The resulting bad press for cameras was met with claims in 2010 by the London Metropolitan Police that they were actually able to solve six crimes a day with camera evidence.[34] Commentators scoffed that most of them were probably jaywalking.

A recent scientific review of various crime prevention techniques looked at thirty-six U.K. studies, ten in the U.S., and one in the Netherlands. These researchers found that "there is little evidence that the following reduce fear of crime: street lighting improvements, closed-circuit television (CCTV), multi-component environmental crime prevention programs, or regeneration programs."[35] Many security cameras are, as Bruce Schneier famously puts it, "security theatre."[36]

A number of motor vehicle registration operations now run an applicant's photo through facial recognition to see if a driver's license has already been issued to the owner of that face. In British Columbia, Canada, the government-run monopoly auto insurer, Insurance Corporation of British Columbia (ICBC), uses facial features including the distance between the eyes as well as cheekbone geometry to root out fraud. Their photo database is of great interest to law enforcement, but there are some thorny privacy issues there.

On June 15th, 2011, downtown Vancouver was engulfed in riots after the home team lost the final game in the Stanley Cup hockey series. People were stabbed, police officers were injured, and there was extensive property damage. Digital devices helped to feed the violence, as rioters reacted to the presence of media and personal cameras. However, digital photos also played a key role in tracking down the offenders.[37]

British Columbia's Information and Privacy Commissioner, Elizabeth Denham, had to decide if using the ICBC's motor vehicles registration database to try to identify offenders would violate the province's privacy laws. She ruled that it was acceptable for the police to provide candidate images to ICBC for possible matching. However, a court order would then be required for the matched-up results to be revealed to the police.[38]

Police in Vancouver also put out an appeal to the public to assist in the massive post-riot investigation. They set up a special website, riot2011.vpd.ca, with photos of "people who are alleged to have committed criminal offences." The public was offered a simple "click and identify" system to provide information.

Did it work? In July 2013, two years after the riot, the Vancouver Police Department (VPD) announced that they had recommended charges against 352 alleged rioters for 1,204 offenses. The accused were as young as fourteen years old. As an example, the report describes three high school friends from Victoria, British Columbia, who "were captured on video committing multiple crimes throughout the night, including break-ins to four separate businesses." VPD Superintendent Dean Robinson says they are not yet finished hunting down suspects, and "those rioters out there that believe they can wait us out and hide with anonymity, we will find you and you will be brought to justice."[39]

There are a number of reasons why the Vancouver Police Department turned to the public for help. One is just good public relations—hooligans trashing the city's downtown does not sit well with most law-abiding citizens. Also, some of the demonstrators had no criminal record, and may have been too young to be in the motor vehicles system.

While the surveillance camera and cell phone videos used by the VPD were certainly helpful, they are nowhere near the state-of-the-art in surveillance camera technology. A remarkable photo of the crowd on West Georgia Street taken a few hours before the riot was posted by a company called Active Computer Services. It is actually a composite image of 216 high-resolution photos stitched together, and it reveals an uncanny level of detail.[40] You can zoom in from the massive scene to identify individual faces with ease. Active Computer Services has a particularly telling motto on their home page ("I spy with my little eye ... ") and they tout the "forensic science" applications of their technology.

According to Charlie Savage in the *New York Times*, scanning for a wanted face in a crowd is still a tough computer science problem.[41] However, Savage writes, progress is being made: the U.S. government–backed Biometric Optical Surveillance System (BOSS) works with two cameras, equipped "with infrared and distance sensors. They take pictures of the same subject from slightly different angles. A computer then processes the images into a '3-D signature' built from data like the ratios between various points on someone's face to be compared against data about faces stored in a watch-list database."

The Department of Homeland Security ran a test of BOSS in September 2013, using it to scan about six thousand fans attending a hockey game in Kennewick, WA. The faces of twenty volunteers were placed in a database. The challenge was to find them among the hockey fans, at a distance of fifty to one hundred meters, quickly enough so that if any were terrorists they could be located and intercepted. The results have not yet been disclosed.

Several commentators have noted that this type of surveillance system is often launched for crime-fighting or anti-terrorism purposes, but people quickly find other uses for it, including commercial ones. The day is not far away when the kid selling soft drinks at a stadium may pass you a note that says your car's lights are on, having linked your face in the crowd to your license plate. They might even figure out a way to charge you for that service.

While this BOSS technology is not yet operational, experts say it will be deployed within five years. Privacy advocates suggest that we need to make rules now about how it can be used in the future, or we will simply default to ubiquitous surveillance.

The FBI, the Department of Homeland Security, and the U.S. State Department are very enthusiastic about facial recognition. According to Brian Merchant, writing for *Motherboard*, "the Department of State currently runs one of the largest facial recognition operations in the world. It uses a database of 75 million photos or so to cross-check visa applications."[42]

While we hear a lot about the prevalence of cameras in the U.S. and the U.K., another country is on track to become the world leader in video surveillance. China is already estimated to have one surveillance camera for every forty-three of its citizens.[43] I have been taken into a secret monitoring center in a major Chinese city where operators watch a gigantic wall of monitors covering every major intersection. The Chinese have a significant home-grown surveillance camera industry, ironically boosted by the United States, which slapped export restrictions on surveillance technology in 1989. That fueled China's own research and development in this field.

Private use of facial recognition technology is also growing daily. The contours of this expansion were neatly summarized by *Motherboard*'s Jordan Keenan, who wrote, "If you use social media, have a driver's license, shop in stores, and walk in public, chances are good that your faceprint will soon be assigned to your identity, and eventually be used on a daily basis to build a profile of you at a level of detail you hoped would never be possible."[44] Improving facial recognition is also the reason you have to maintain such a stern expression for your passport and visa photos.

On August 8, 2000, a woman wearing a toque and dark sunglasses entered a Safeway store near downtown Calgary, Alberta, Canada, pushing a toddler in her shopping cart. She wrote a note addressed "To Whoever finds my son," wheeled him into the cookie aisle, and simply left the store. According to media reports, the two-year-old kept saying "where is my Mommy?" but she was nowhere to be found.[45]

Police were called, and appealed for the public's assistance on the television news. When the mother did not come forward, Alberta's Minister of Children's Services ordered publication of this photo, shown here at reduced quality to preserve privacy.

Figure 5. Woman in Safeway with baby. Courtesy of Government of Alberta.

The baby's mother was soon tracked down in the State of Washington. But how was Safeway able to supply that picture? I visited the store and found the inconspicuous camera posted over the entrance. Sure enough, everyone who entered was being captured on video.

Back in 2000, this was a shocking discovery for me. It seemed unnecessary for a grocery store to capture the arrival and departure of every customer on video. What else were they watching? Now, it is hard to imagine an urban space that is not within the reach of a surveillance camera. It's not just that they're capturing your image: it's what might they might do with it, now and in the future.

Just what are all those surveillance cameras doing when they are not taking pictures of suspected terrorists, shoplifters, or mothers who abandon their kids? They are watching us, creating a potentially eternal archive of everything we do. The same technology that allows law enforcement to zoom in on bad guys can impinge on the privacy of law-abiding citizens in some very creepy ways.

The one-way nature of surveillance cameras is one of their most unsettling features. Aside from the occasional blinking light, they tell us nothing. We tell them everything. One way to level this playing field is to wear our own cameras. University of Toronto Professor Steve Mann coined the term "sous-veillance" to describe the countermeasure of wearing cameras to record our own version of how things happen.

One of the first famous uses of this approach was the 1991 videotaping of the beating of African-American construction worker Rodney King by the Los Angeles police. The police officers were acquitted, despite compelling video evidence against them, sparking the 1992 riots in that city.

Now, dashboard cameras are commonplace, at least in the United States, and definitely in Russia, where they are almost mandatory to survive the country's traffic and scam artists who stage fake accidents.[46] Video evidence gives you an edge in many situations, and some people are already logging their lives as an offbeat kind of hobby. I spoke to one of these lifeloggers, and he estimates the cost of recording his every moment in audio and video at about one dollar a day for storage media. His cost in terms of relaxed social interaction, however, might be much greater. The possibility of recording everything you see, hear, smell, and touch was also the subject of a research project called LifeLog, funded by the Defense Advanced Research Projects Agency in 2003 but abruptly canceled in 2004 after privacy groups voiced objections.[47] According to many experts, the program has continued, at least in spirit, both inside and outside the U.S. government.[48]

The movie *Déjà Vu* (2006) envisioned a world in which satellites look down at people and peer inside their homes with laser imaging, using computer reconstruction to replay a terrorist attack and then travel back in time to avert it. Camera technology is definitely moving in the creepy direction suggested by the movie. Scientists at MIT have announced the ability to see through solid walls to an accuracy of ten centimeters. PhD student Fadil Adib, speaking on a *Network World* video, says "we're doing localization through a wall, without requiring you to hold any transmitter or receiver, simply by using reflections off the human body."[49]

The researchers gave their project a benign name, "Kinect of the Future," suggesting it might simply be the next evolution of Microsoft's popular gaming device. However, a system that can peer through walls will have applications far beyond video gaming. It probably would have been greeted much differently if they had called it the "Anne Frank Finder." That may indeed be much closer to how this technology will really be used.

Even before the Boston Marathon bombings, the U.S. Air Force contracted with the 3D biometric imaging firm Photon-X for a new kind of surveillance camera. By using a combination of infrared and visible light, and by indexing muscle movements that are unique to each individual, the company claims it can produce a unique "bio-signature" for a person and then silently track them.[50]

The company is also promoting something they call the Spatial Phase Imaging Technique (yielding an unfortunate acronym, SPIT), which purports to read your fingerprints at a distance of up to ten feet, with "longer distances being developed." They also claim they can "passively capture 3D geometry for skin, hair, eyes, teeth, clothing, and anything else that is in frame, with no special preparation of the subject."[51]

While surveillance cameras do not yet follow us everywhere, we do a pretty good job of filling in the gaps with our own cameras. We snap billions of photos and many of them end up on Facebook and photo-sharing websites. By putting our real name next to photos, we provide the fodder for all kinds of nefarious data mining.

TV studio cameras have large red "tally lights" to show the anchorperson where to look, but far too many unwitting TV presenters have been embarrassed by their "off camera" comments that made it to air, so they don't really trust the lights.

While their lenses may be almost invisible, laptop computers and smartphones are equally risky. Unless you douse it in a glass of water, as a friend of mine did when he learned his smartphone was infected with some nasty malware, there is a decent chance that your camera can be hijacked by a hacker.[52]

Showing off clever ways to remotely invade a smartphone has been a staple of hacker conferences for years. Now, you do not even need hacker skills to take over someone's smartphone, because "there's an app for that"—in fact, many of them. One is the notorious "Rastreador de Namorado" (Boyfriend Tracker) from Brazil.[53] Once you slip this onto someone's phone, it reports all of the device's travels logged by the phone's GPS. It reveals steamy text messages sent to other lovers. It even allows the scorned paramour to call the phone, and put it into audio eavesdropping mode.

People are coming around to the realization that our smartphones may indeed be traitors, but at least we can trust home appliances like our television sets. Or can we?

Seungjin "Beist" Lee shuffles nervously as he stands in front of the ragtag press room contingent at the Black Hat USA 2013 conference in Las Vegas. Seasoned business journalists mix with earnest young tech bloggers and local TV reporters awkwardly trying to explain hacking to the general public. It is not the easiest of tasks, because many of the exploits and vulnerabilities on display are pretty technical. Mr. Lee's presentation, however, is understandable and frightening to everyone.

"Usually at home people have their TVs in the bedroom," he says, "because you can watch the TV well." OK, no surprises there. Reporters continue to doze and check email.

"On the other hand, the TV can also watch you very well." Reporters look up. What did he just say?

"I want to say I hardly wear clothes at home, I just wear underwear." Where is he going with this?

"But I don't care about it, but I care about my family and my girlfriend." Ahh, he's saying that hackers can remotely activate the camera in those new high tech TVs. This *is* interesting.

Lee attempts to be coy about just whose Smart TVs he is talking about, since the company has now paid him a consulting fee; but the tenacious press corps relentlessly drags it out of him. We already know he is Korean and he drops the big hint that "the company starts with an S."

In his presentation, Lee promises he will not only demonstrate the technical hack he discovered; he will also show how "Smart TVs monitor you 24/7 even though users turn off their TV, meaning #1984 could be done." It is interesting, and somewhat chilling, that a young Korean graduate student would remind us of George Orwell's dystopic world with a Twitter-style hashtag.

Lee can activate the camera remotely because a Smart TV is not just a TV. It is really a computer, microphone, and digital video camera wolf hiding in the sheep's clothing of a familiar household appliance. Some reporters wonder whether anyone besides a stay-up-all-night hacker would bother exploiting vulnerability like this. The answer is a resounding "yes!", because once the secret is posted online, "script kiddies" all over the world can start using it without even understanding how it works, often with tragic results.

Far too many people, of both genders, have committed suicide after having their intimate photographs posted online. Sometimes the pictures were given voluntarily to a "friend" who cruelly shared them with a wider audience. In other cases, acts of sexual brutality have been videotaped and posted for all to see. The most common scenario now appears to be "sextortion," where the malefactor obtains some compromising photos, then demands more.

Two Canadian teenagers, Amanda Todd and Rehtaeh Parsons, were driven to suicide by online sexual harassment and bullying in 2012 and 2013, respectively. The Canadian government responded by introducing legislation to make it a crime, punishable by up to five years in jail, to distribute "intimate images" without consent. In the U.S., according to an article in *ABA Journal*, "only two states, California and New Jersey, make it illegal to post a sexual photo online without the subject's consent."[54]

Cassidy Wolf, Miss Teen USA 2013, was the victim of a webcam-enabled sextortion attack in which the perpetrator "used malicious software and tools to disguise his identity in order to capture nude photos or videos of female victims through remote operation of their web cams without their consent."[55]

The FBI press release on the case goes on to say that the nineteen-year-old perpetrator, Jared James Abrahams, "threatened to publicly post compromising photos or video to the victims' online social media accounts, unless the victim either sent nude photos or videos, or engaged in a Skype session with him and did what he said for five minutes." Abrahams was sentenced to 18 months in a federal prison after pleading guilty.

Miss Wolf has done a huge, if unwilling, service by bringing this type of attack into the public's consciousness. While the motivation here seems to be of a sexual, not commercial, nature, that did not stop companies from finding a way to take advantage of it. A Google search on "Cassidy Wolf" produces stories from mainstream news media such as CNN and the *Los Angeles Times*. However, many of the other "hits" are in fact rewrites of those mainstream news stories sponsored by a company that makes, you guessed it, lens covers for web cams and Smart TVs. Using aggressive search engine optimization, this company managed to insert its "Protect yourself now with a web cam cover by ... " message into the conversation about Cassidy Wolf.

The commercial exploitation of web cameras took a frightening, and definitely illegal turn in a case involving rented computer equipment. If you were unfortunate enough to lease a computer from one of seven U.S. firms, or their international affiliates, you also received hidden software called "Detective Mode." According to the U.S. Federal Trade Commission, "when Detective Mode was activated, the software could log key strokes, capture screen shots and take photographs using a computer's web cam."[56] The software also contained a "kill switch" which could disable the computer if it was stolen or, more commonly, if the renter fell behind on the payments.

The FTC noted that "using Detective Mode revealed private and confidential details about computer users, such as user names and passwords for email accounts, social media websites, and financial institutions; Social Security numbers; medical records; private emails to doctors; bank and credit card statements; and web cam pictures of children, partially undressed individuals, and intimate activities at home."

The *American Banking Association Journal* reported that the retailers agreed "to restrict the use of PC Rental Agent software developed by Pennsylvania-based DesignWare that previously allowed more than 1,600 licensed rent-to-own stores in the United States, Canada and Australia to spy on over 400,000 customers."[57]

Remarkably, the rental companies escaped without a fine. A school district in suburban Philadelphia was not so fortunate. It was accused in a lawsuit of using a similar technology, the now defunct "TheftTrack," to spy on students in their homes. In the class action lawsuit, students alleged that school officials were activating the cameras in the computers while they were off school premises. In a tremendous display of either arrogance or stupidity, school officials actually disciplined one student using photos taken surreptitiously in his bedroom as evidence. The judge awarded that student $175,000 in damages.[58]

Most people agreed that these cases represented a creepy use of technology since the users were unaware that they were being watched. However, some people must consent to video surveillance as a condition of employment. Jobs including bank teller, bartender, day care worker, and even zookeeper come with the expectation that you will be watched to make sure that you are not stealing from the company or doing something even worse. Ordinary office workers have generally been immune to this expectation, though that is changing. For instance, it is now possible to spy electronically on government functionaries in the Office of the Chief Minister in Kerala, India. Anyone with Internet access can see who is snoozing on the job, or has gone off for chai, or, heaven forbid, is accepting a bribe.[59]

Do people actually look at workplace cameras? If we believe the visitor counter at Dental Office K in Aomori Prefecture, Japan, "the first WEBCAM of [a] dental office in the world," over half a million visitors have taken a peek since that office began streaming live images in 1996.[60]

We may all be peering into a video portal at work if the folks at Vancouver-based Perch Communications are successful in marketing their "always on video portal." In an interview with CBC Radio's Nora Young, CEO Danny Robinson likens it to having a window into a co-worker's office, except that "the people might be 3500 miles away."[61] While the video is continuous, the audio is off until you walk up and look into the device. It then activates the microphone until it sees you walk away. Even Robinson says that having this thing sitting on your desk is "just a little bit on the freaky side for most people," and he suggests putting it in a hallway.

Confirmation that walls that watch us may soon be commonplace comes from a fascinating if rather creepy experiment from

Washington, D.C.-based iStrategyLab. Their S.E.L.F.I.E. ("Self Enhancing Live Feed Image Engine") is a two-way mirror with a camera and computer mounted behind it. As they explain on their website, the device is "triggered by simply standing in front of the mirror and holding a smile." When it sees you are at your smiley best, the device initiaties a countdown, then uses LEDs to simulate a 'flash' as your photo is taken. The resulting image is posted directly to your social media feed.[62]

Human-computer interaction experts agree that "always on" video links will take some getting used to, and people may find them disturbing, especially if they are installed by the boss. They are, in an uncanny valley sense, human (because human eyes and ears are processing the images and sound remotely) and simultaneously non-human since they are devices hanging on the wall.

Many police forces are equipping their members with body-worn cameras to document arrests and other interactions with civilians. A year-long study in Rialto, CA found that the cameras resulted in "more than a 50% reduction in the total number of incidents of use-of-force compared to control-conditions."[63] The authors suggest that the behavior of citizens may also have been modified: "Members of the public with whom the officers communicated were also aware of being videotaped and therefore were likely to be cognizant that they ought to act cooperatively." One of the study's authors was Rialto's police chief, William Farrar. He reports there was some reluctance by his officers to wear the cameras, which they referred to as "Big Brother." However, as Chief Farrar told the *New York Times*, he reminded them that civilians can use their own smartphone camera, "so instead of relying on somebody else's partial picture of what occurred, why not have your own?"[64]

While the presence of cameras can make people more civilized, it can also have the opposite effect. Stories abound of bystanders pointing their smartphones at accident scenes and rapes in progress instead of calling for assistance. The Vancouver police blame cameras for fueling some of the violent rampages in that city, commenting that "the 2011 riot can be distinguished as perhaps the first North American social media sports riot."[65] The police report goes on to say that "the acting out for the cameras seen in the 1994 riot was multiplied many times more in the 2011 riot by the thousands of people cheering the rioters on and recording the riot with handheld cameras and phones."

Cameras have come a long way since Matthew Brady captured the horrors of the Civil War on glass plates in the 1860s. They have morphed from a heavy object that sat on a substantial tripod and needed multi-minute exposures into chip-sized sensors that fit in our smartphones, laptops, even a pair of glasses. The next step is quite likely to be a camera in a contact lens. Korean researchers have created a proof of concept of this and tested it on rabbits.[66] Mounting them on everything from traffic lights to dashboards to police officers changes cameras from something that we pull out, deliberately aim, and focus into an organic, omnipresent part of our environment.

Just as GPS chips became smaller and cheaper and are now installed in your smartphones and camera and perhaps soon in your newborn baby, camera chips will continue to proliferate. The 3D printing revolution will allow people to make a hollowed-out button and stick a tiny camera inside to sneakily capture your photo.

Then again, they might just wink at you while wearing Google Glass.

# Image Creep

I saw Google Glass before it was even a twinkle in Eric Schmidt's eye.

As a technology writer and reviewer, I was sent demo versions of all sorts of products, including some that never made it to market. In the mid-1980s, a package arrived with one of the first heads-up television displays aimed at the consumer market. It was a set of glasses with a tiny monitor and a prism that allowed you to watch TV while still participating in normal life.

The device, now consigned to the tech dustbin, did give me one moment of profound technocreepiness. I was testing it one night in my university office, using it to watch *60 Minutes*. The cleaning lady came in to empty the trash. I will never forget what happened next. I saw a chimera—an elderly lady's body with Mike Wallace's head grafted on top. I screamed. She screamed. It seemed like a dumb way to watch TV, so I sent the thing back and wrote a lukewarm review: it was also extremely uncomfortable to wear.

The introduction of Google Glass has brought this type of technology literally to the public's eye. All of a sudden, people are walking around with a device that enhances their ability to grab information out of the ether. Google Glass wearers can potentially recognize your face as they shake your hand, and then casually glance upwards to retrieve your kids' names and birthdays.

But what really alarms many is that Google Glass can also secretly take a picture, or record a video, and immediately upload it to the Internet, just by the wink of an eye or the raising of an eyebrow. Google Glass does have a light to indicate when it is taking a photo or recording video. People promptly found ways to subvert it.

Chris Barrett, one of the earliest users of Google Glass, said he was having trouble using the device in bright sunlight. So, he designed a clip-on sunshade, a piece of plastic that can be run off on a 3D printer. He has even made the code for it freely available online.[67] In addition to blocking the sun, Barrett's creation happens to cover that pesky light that tells others you are recording them.

Barrett loves wearing his high tech eyewear in places it is not supposed to go. He has reportedly filmed inside an Atlantic City casino and, in a stroke of luck, apparently became the first person to record an arrest with Google Glass. This little documentary was quickly posted to YouTube.[68]

Dozens of innocent bystanders, some of them children, appear in that video, and many are facing the camera. Coupled with massive facial image databanks, and advances in facial recognition software, they could probably be identified. A bizarre new social rule is emerging: if you are really trying to protect your privacy, you should stay away from arrests, car accidents, riots, and landmarks—anything that people are likely to photograph. Perhaps you should not go out at all.

The camera function of Google Glass and similar devices is really a giant social experiment to redefine where photography is acceptable and what behaviors will get you called a creepy "glasshole," a cheeky term that even Google has started using.[69] Even in a public place, where photos are generally fair game, people have reported being very disturbed by strangers taking their photos; especially pictures of their children.

Gym locker rooms have long been off limits for cameras. Cell phones are often now banned there, too. While I was embedded with the Canadian Forces in Afghanistan, my cell phone was confiscated for fear that I would take a photograph, perhaps even a geo-tagged one, of something I should not. Casinos have always banned cameras, except for the fleet that they operate themselves. Their latest challenge is gamblers with tiny cameras hidden in their sleeves to watch the cards as they are dealt.[70] Google Glass is already forbidden at some sports arenas, movie theaters, and concert venues. Hospitals are implementing bans, as will many other workplaces. Even hotel lobbies may become off limits.

While working as a TV journalist, I was asked to leave a posh hotel while trying to shoot a "stand up" in the lobby. The manager explained that "we need to protect the privacy of our guests—you just might catch somebody here with his mistress or something like that." Legally, they have the right to do that in most jurisdictions.

Internet Rule 34 ("If it exists, there's porn of it") strikes here with vengeance. Soon after Google Glass became available, the "first Google Glass porn" appeared online at the adult app store mikandi .com. Their inaugural offering features a cameo appearance by Ron Jeremy, holder of the *Guinness Book of World Records* title for "Most Appearances in Adult Films." Now a senior citizen, he keeps all his clothes on in this short piece of point-of-view pornography. A censored version is available on YouTube.[71]

There seem to be no limits to human stupidity when it comes to posting inappropriate images online. People have distributed photos of themselves in an unbuttoned airline stewardess uniform (Ellen Simonetti, the "Queen of the Sky" blogger who flew for Delta airlines) and taking a bath in the sink at Burger King (Timothy Tackett).[72] Both were fired from their jobs, though they went on to other careers, propelled no doubt by the notoriety from their online misadventures.

Simonetti and Tackett are consenting adults, but things are quite different when children are involved. Often they seem either unaware of or unconcerned about the risks of releasing questionable comments and images into the Wild West of cyberspace. In a CNN report called "The Secret Life of My Sixth Grader," a mother creeps herself out by spying on the texts and Instagram photos of her eleven-year-old son, who, she notes, "has never let on that he is remotely interested in girls."[73] The content of his messages and photographs soon convinces her otherwise. "Maybe it's the digital photo-filters," she muses, "but the girls seem sexy beyond their years."

A 2008 survey by the National Campaign to Prevent Teen and Unplanned Pregnancy asked teens aged thirteen to nineteen and young people aged twenty to twenty-six if they had ever posted or sent nude or semi-nude photos of themselves. Overall, 20% of the teens and 33% of the young adults answered in the affirmative, with more females than males admitting to this activity in both age groups.[74]

Skype video chat, and sites like Chatroulette, created by a seventeen-year-old Russian lad, have brought inappropriate images into the video domain. Turn on your web camera and microphone and meet new friends from around the world, chosen for you randomly. They may or may not be wearing clothes.

In 2011, researchers working closely with Chatroulette introduced a filtering technology to try to cut down on the surprise of sudden, unwanted nudity. According to one report, the "filter technology and moderation (by human censors) results in the banning of 50,000 inappropriate users daily."[75]

Bare skin-finding technology works both ways, of course, and there are now apps that will mine photos posted online specifically for nudity or some close approximation. As just one example, Badabing!, available in the iTunes stores, claims it will save you the effort of browsing "endlessly through a friend's albums looking for beach or pool pictures."[76] The developers call it "the only image social recognition app," but it certainly won't hold that title for long.

The wildly popular smartphone app Snapchat would appear to address the problem of persistent photos, since images sent through it disappear within a few seconds. However, this may well be a false sense of security. Armed with the right forensic tools, experts have been able to mine smartphones for supposedly deleted Snapchat photos. There are also apps like SnapSave and SnapChat Save Pics that are explicitly designed to defeat the ephemeral aspect of Snapchat photos and videos. These apps work around the Snapchat system, so, as one says, "the sender of the snaps doesn't get notified" that you are not playing by the rules.[77]

Even without these apps, someone can always point a camera or another smartphone at a fleeting Snapchat image and capture it for posterity. Just assume that if you send a photo, it can be grabbed and have a life that extends far beyond momentary ogling.

Increasingly, your face is becoming a key that unlocks a vast amount of personal information about you. This was brought to the public's attention when some users of a leading U.S. matchmaking site learned that their online personas were not as private as they thought they were.

Dating site users almost always include a photo, but typically register using a pseudonym, like sexybaby235 or hungguy404. The would-be lovers only reveal their true identity when someone of interest comes along. Carnegie Mellon University professor Alessandro Acquisti grabbed almost six thousand dating site profiles from Match.com and compared them against publicly available Facebook profile photos in the same geographical area using a facial recognition program called PittPatt. His goal was to "de-anonymize" people.

He immediately encountered an interesting research problem: just because the computer says two faces match up does not mean it is true. The human brain is still the world's best facial recognition engine, so Acquisti enlisted human reviewers to rate the computer's work. Amazon runs a business called Mechanical Turk. Its name pays homage to the mysterious chess-playing robot constructed in Europe in the late 18th century that secretly contained a chess-playing human being. Amazon's system is a high tech "piecework environment" where people agree to do menial tasks, like looking up a company's main office address, for a few pennies each. If you do enough of these, fast enough, you can actually earn some decent cash.

The researchers asked the Mechanical Turk validators to sort the computer's proposed facial matches into categories like "sure match" and "highly likely." Acquisti's conclusion was that "about one out of ten dating site's pseudonymous members is identifiable."[78]

He freely admits that "no human being can really take the time of having one browser open on Facebook and the other browser open on the dating site, and hope to find matches." However, as this experiment demonstrates, face matching can be automated. This project involved comparing more than 500 million pairs of faces, which would take any of us a pretty long time. A computer can do that in a flash and come back asking for more.

To illustrate how our faces are becoming excellent personal identifiers, he performed another study, this time on a U.S. university campus: "Passers-by were invited to participate in the experiment by sitting in front of a webcam for the time necessary to take three photos, and then by completing a short survey. While a participant was completing her survey, her photos were uploaded to a computing cluster and matched against a database of images from profiles on the social networking site. Thereafter, the participant was presented with the images that the facial recognizer had ranked as the most likely matches for her photograph. The participant was asked to complete the survey by indicating whether or not she recognized herself in each of the images. Using this method we re-identified a significant proportion of participants."

Acquisti expects that you will soon be able to point a smartphone at someone and learn quite a bit about them in real time. Speaking at the TEDGlobal 2013 conference, he said: "Pushed to an extreme, you can imagine a future with strangers looking at you through Google Glass or their contact lens, and with seven or eight data points about you they could infer anything else about you."[79]

Professor Acquisti is a scientist, but this field is also a playground for pranksters and people trying to make a point about online privacy. A guerrilla filmmaker named Jack Vale set out to baffle total strangers by showing how much he knew about them. First, he says, he searched for "Twitter, Instagram and other social media posts close to my current location."[80] He then walked up to people in those posts with comments like "Is your name Jessica?" or, with devilish accuracy in one case, "I just wanted to wish you Happy Birthday." He often knew where they worked and the names of their pets. His subjects found it extremely creepy.

In an even more grandiose demonstration of online privacy risks, the Belgian Financial Sector Federation set up a tent in a square in Brussels and invited people to be part of a TV program with a "gifted clairvoyant named Dave." After some theatrics like hugging them and jumping around, he proceeded to tell them their most intimate details, from hidden tattoos and secret sexual preferences to their bank account numbers and precise balances.[81]

At a strategic moment, a curtain drops to reveal hard working hackers dressed in black, bringing up the subject's social media pages on large computer screens and feeding the information to Dave. Instead of a psychic TV show, participants became part of a public service announcement about the risks of sharing too much information online. The tagline is: "Your entire life is online. And it might be used against you."

Posting photos of friends and family has become a favorite recreational activity for many people. Good natured sites like awkwardfamilyphotos.com provide hours of amusement and are probably just fine if the subjects in the photos have consented to be there. But sometimes, personal photos are simply appropriated for commercial and even nefarious purposes.

Tennessee parents Pamela and Bernard Holland were shocked to see an image of their son, Adam, who has Down syndrome, on a website. To their horror, and without their knowledge or consent, a radio station posted a digitally altered photograph of their son holding a sign that said "Retarded News." While the station apologized, the parents' $18 million lawsuit demonstrates that expropriating and publishing a digital photograph can have serious consequences.[82]

A similar creepy shock awaited Scranton, PA mother Kaylee Doran who found that photos of her baby that she posted on Instagram were being "re-posted with insulting comments about her son" calling him "ugly" and "disabled."[83]

Which brings us to the disturbing online world of "baby role playing." There is a thriving underground online culture of "fantasy adoptions" and caring for virtual babies, and even children and teenagers. There are often clear sexual overtones, with comments such as "all our babies are breast fed" and unsettling instructions about how to change your teenager's diaper.

A mother from Hamilton, Ontario was shocked to find a photo of her baby daughter listed as "up for adoption" on one such website. The outraged mother responded with "Uh, I'm sorry ... but this is definitely NOT 'Ally' and she is definitely NOT for freaking sale! This is MY child, and I did NOT and would NEVER give permission for this post!"[84]

The photos used on the fantasy adoption role playing sites do not generally meet the legal definition of child pornography, but they are highly unsettling. The scope of this and other online sexual fetish subcultures is difficult to estimate. However, one site alone, FetLife.com, boasts about 2.8 million members with almost 15 million pictures posted. In addition to having, as they put it, "a fetish about security," the site assures prospective members that "Your kinky friends are already on here."

FetLife is an example of the "Deep Web"—a vast section of the Internet that is not indexed by common search engines like Google. According to some estimates, 99% of online content is out of reach of the search engine spiders and the casual user.[85] This includes huge databases that are accessed by entering specific queries, proprietary content behind paywalls, and data on corporate and university Intranets.

There is also a whole array of deliberately hidden information, much of it illegal, which is where some of the most disturbing Internet images are hidden. In a report on the Deep Web, CNN noted that many of these sites use Tor (The Onion Router) to further obfuscate what they are doing. According to the report, secret websites that end in .onion offer "stolen credit cards, illegal pornography, pirated media and more. You can even hire assassins."[86] While the most famous online black market, The Silk Road, was shut down by law enforcement on October 2, 2013, new versions are popping up.[87]

The legality of using private photos without permission is still being worked out in the courts, and the law depends on where you are located. Europe is very concerned about privacy issues, and even in California, there is a growing awareness that your photos should not simply be there for the taking.

A San Francisco judge ordered Facebook to pay a total of nine million US dollars to 600,000 Facebook users whose photos were used without their consent next to "Sponsored Stories" advertisements. Facebook promptly changed its data use policy so it could continue the practice without legal risk, prompting *Information Week* editor-at-large Thomas Claburn to observe that "Facebook Says User Data Is Price of Admission."[88]

In a similar fashion, Google's Street View has revealed people engaging in activities, such as leaving the adult video store with a stack of tapes, that they would rather keep private. Yet for every photo that is grabbed without permission, thousands are posted by willing users who are eager to share their lives with their Facebook "friends." Those of us who are not celebrities or politicians may feel that our photos are not interesting enough to be valuable. However, personal photos can be monetized in some rather creepy ways.

In many U.S. states, if you are arrested, the details of your alleged offense, accompanied by your mugshot, will be placed online by local law enforcement. Non-Americans often find this startling and a bit unsettling, but the U.S. Supreme Court has affirmed that such posting is legal. Why would a police agency go to the trouble of posting mugshots on its website? One justification is to inform the public about alleged wrongdoers in the community. A more likely explanation is that sheriff is an elected office in much of the U.S. The best way to show you are doing your crime-fighting job is to have a steady stream of seedy-looking "just arrested" mugshots gracing your website, right next to your own smiling face in uniform.

Mugshot photos are generally posted online when a person is arrested, long before there is any determination of guilt. While frequently deleted after thirty days, sometimes the data sticks around. For example, the Pinellas County Sheriff's Office in Florida has arrests going back to November 2005. Some people appear to have their photos posted there for relatively trivial offenses, such as "trespassing in a park" which is probably legalese for "being homeless."

The images are not censored and the full legal names and aliases of offenders are provided. I even found one sheriff's office showing a child's full name and photo.

We do not know if little Bobby (age twelve years, two months) was guilty, but we do know that he was publicly humiliated on the Internet, which for many young people is the most potent form of punishment. It is true that the record of his 2010 arrest is probably long gone from the Sheriff's website. But the very fact that it is captured and included here illustrates that "on the Internet things never, ever completely go away."[89]

Even foreign nationals can be swept up in the mugshot dragnet. A Fort McMurray, Alberta, legislator was arrested in Minnesota and charged with one count of "hiring, offering to hire or agreeing to hire a prostitute in a public place." The Minnesota law enforcement agency dutifully posted the full and embarrassing arrest details, complete with his full name and home address back in Canada.

As if this posting of mugshots by law enforcement were not invasive enough, a thriving, for-profit "mugshot industry" has sprung up. Their websites copy images from law enforcement sources, giving them much wider exposure, often for a longer period of time. This voyeuristic phenomenon started with celebrities such as Bill Gates and Lindsay Lohan, but now just about anybody's image could grace a site like mugshots.com.

This website states that its mission is "to inform the public of arrests and hold government accountable," though those noble-sounding principles appear to be exercised with some flexibility. On their home page the company offers to "unpublish" your photo for a fee. Bizarrely, they also offer to "permanently publish" a particular photo if you really dislike somebody or are proud of your own crime. It is not clear what happens if they get conflicting publish and unpublish requests and payments for the same photo.

Sensing a class action opportunity, a U.S. group, classaction-againstmugshotwebsites.com, is now raising funds to get mugshot posting sites banned. Several states, including Oregon, have already passed laws regulating these sites. That state's version requires website operators to take down an image for free if you are not convicted of the crime or the charge is downgraded to a violation. The site owners have thirty days to comply, which does not help you much if you have a job interview the following week. And, of course, your scowling or grinning mugshot may have already been grabbed and gone viral, especially if your name is Justin Bieber.

The mugshot industry is an excellent example of how opportunistic business ventures can spring up *ex nihilo* when new technologies enable them. The mugshots were always around in dusty filing cabinets, but the advent of easy photo sharing and search engines such as Google have turned viewing mugshots from a pastime for the very weird to a mainstream activity. While legal measures like Oregon's can be helpful, there will always be rogue sites like those hosted offshore, out of reach of law enforcement.

This is why experts believe that the best way to fight creepy uses of technology like this is to strike back with technology. That is exactly what is happening. Google has already announced that it will be demoting mugshot search results so that one youthful mistake does not pop up above decades of good works and community service. PayPal and the credit card companies are also looking at limiting the ability of mugshot sites to take money, which would probably have more impact than any law since this industry is driven by credit card payments.

When Wikileaks was faced with a similar financial blockade, it turned to the anonymous currency Bitcoin. Mugshot operators might wind up telling people who want their pictures removed to pay in Bitcoins, or wire them money through Western Union.

As Kashmir Hill points out in *Forbes*, "Private industry may wind up doing what lawmakers are constitutionally forbidden to do: killing an ugly information practice by both burying it in search results and cutting off its funding sources."[90] Even if the stream of mugshots dries up, malefactors will still have plenty of gold to mine from postings on Tumblr, Instagram, Flickr, and Facebook.

Not everyone who sends you a friend request on Facebook is necessarily your friend. In fact, Facebook's dark side goes back to its very creation. The digital world is a wonderful preserver of information, both wanted and unwanted. Courtesy of a blog entry that was produced as evidence in a court case, we have some verbatim insight into Mark Zuckerberg's thinking on the very night he came up with what would become Facebook:

*now I just need an idea ...*

*9:48pm. I'm a little intoxicated, not gonna lie. So what if it's not even 10pm and it's a Tuesday night? What? The Kirkland facebook is open on my computer desktop and some of these people have pretty horrendous facebook pics.*

*I almost want to put some of these faces next to pictures of farm animals and have people vote on which is more attractive. It's not such a great idea and probably not even funny, but Billy comes up with the idea of comparing two people from the facebook, and only sometimes putting a farm animal in there. Good call Mr. Olson! I think he's onto something.*

*11:09pm. Yea, it's on. I'm not exactly sure how the farm animals are going to fit into this whole thing (you can't really ever be sure with farm animals ... ), but I like the idea of comparing two people together. It gives the whole thing a very Turing feel, since people's ratings of the pictures will be more implicit than, say, choosing a number to represent each person's hotness like they do on hotornot.com. The other thing we're going to need is a lot of pictures. Unfortunately, Harvard doesn't keep a public centralized facebook so I'm going to have to get all the images from the individual houses that people are in. And that means no freshman pictures ... drats.*[91]

What emerged from Mark Zuckerberg's rather crude ramblings that night was:

- a privacy reprimand by Harvard officials
- a website called Facemash that became Facebook
- a company with a book value over fifteen billion dollars
- a communications system that has 1.3 billion active users just about everywhere on the planet, some of whom have to defy their repressive national governments to sneak in a posting or two

I can assure you that Mark Zuckerberg did *not* create the fundamental concept behind a face book. I know this because I am holding a dusty "1970 Freshman Directory" from Columbia University. Long before Mr. Zuckerberg was even imagined, college students were already judging each other and ridiculing awkward high school grad photos, in dorms and dining halls across the country.

One of the Columbia College men would trade a copy of our all-male book with a student from the all-female Barnard College just across Broadway, and we would engage in the same late-night "hot-or-not" discussions that Zuckerberg automated. However, things posted on Facebook now travel instantaneously around the world, which vastly increases the potential impact on our lives.

In 2010, Facebook introduced a new feature to automatically tag people in photos through facial recognition. This "tag suggestion" feature was turned on by default, a situation that did not sit well with data protection authorities, especially in Europe. The people there seem to be a lot keener on privacy protection than many other nationalities: images from the 1940s of punch cards with meticulously-typed

Jewish names are etched into the public's consciousness. Remember that the vast majority of Facebook users provide their real names and photos, bowing to the company's terms of service.

When data protection commissioners in both Ireland and Hamburg objected to automated facial recognition, Facebook removed tag suggestions from customers in those countries.[92] In fact, they even removed the feature for users in the U.S. for a while, though it has been brought back in substantially the same form.[93] The company is coy about the exact number of photos that are in its database, but did say in an SEC filing that "on average, more than 250 million photos per day were uploaded to Facebook in the three months ended December 31, 2011."[94] So, Facebook gets to build the world's largest, self-validated photo database on the planet, a project which has mind-boggling value for everyone from marketers to dictators to law enforcement agencies.

Not content to rely on careless, lazy humans to properly tag photos with names, Facebook's Artificial Intelligence Group in Menlo Park, CA has been hard at work on "DeepFace: Closing the Gap to Human-Level Performance in Face Verification." In a 2014 academic paper they reported that their method "reaches an accuracy of 97.25% on the Labeled Faces in the Wild (LFW) dataset."[95] LFW, maintained at the University of Massachusetts, is a popular collection of more than thirteen thousand faces with names attached that is often used for testing facial recognition technology.[96] This amazing performance rivals that of humans, who, we are told, are only about one quarter of a percent better (97.53%) than the algorithm. The scientists accomplished this by training a neural network "on an identity labeled database of four million facial images and by applying 3D rotations to align images."

The announcement of DeepFace was greeted with headlines such as "Just as Creepy as It Sounds"[97] and "Facebook's Freaky DeepFace Program Knows Your Friends Better Than You Do."[98] The technology is still on the drawing board but it is hard to imagine it will not move into mainstream use quickly, perhaps even built in to your next smartphone.

The *Onion News Network* has a tongue-in-cheek video report claiming that Facebook is actually a "massive online surveillance program run by the CIA." It goes on to say that "Facebook has replaced almost every other CIA information gathering program since it was launched in 2004." The report praises "CIA Agent Mark Zuckerberg, who runs the day to day Facebook operation for the agency." It jokes that Facebook's Calendar feature even shows where you will be in advance so "now if they want to pick you up for questioning, all they have to do is see which events you've RSVPd 'Yes' to."[99]

In the same vein, a conspiracy theory video on YouTube called "Does what happens in the Facebook stay in the Facebook?" tracks some of the early backers and funders of Facebook, highlighting their defense and intelligence community connections.[100]

I once put the question "Did the CIA create the Facebook" to someone senior enough in that agency to have a well-informed opinion. "No, we did not," he said, but then he added that they use it every day as an excellent source of intelligence, and if it had not been launched by Zuckerberg, the CIA might well have created something like it.

MIT professor and author Sherry Turkle explains why we have such a burning desire to share our lives with the online world, including total strangers, in her book *Alone Together.* Turkle tells us that interacting with machines "may offer the illusion of companionship without the demands of friendship."[101] Although she is writing about robots, the point applies to Facebook as well. Although your ultimate goal may be to communicate with other human beings, when you add something to your Facebook you are in fact dealing with a technological entity.

Blogger Margie Clayman expands on that idea, suggesting that "perhaps people share pictures of them[selves] with their children because they feel a need to prove that they spend enough time with their children. Perhaps people post pictures of new purchases or great meals because they want to prove that their lives are really good."[102]

Photos of our favorite meals and recent purchases have marketers licking their chops. Analysis of that data, combined with images in our postings, can easily reveal other things that we might want to buy. To illustrate how this might work, consider a photo of loved ones that I posted, and tagged, on Facebook.

In the very near future, some computer will probably analyze this snapshot and come back with: "Hmm, (tag: Keri) that poor dog's (tag: Joey) leash is looking rather tattered—we just happen to have a sale on them, running through today only. And those boots you seem to like, we've got them too. By the way, that ski resort you seem to enjoy still has space available over the Christmas holidays—if you book now. Click here. Oh, and are you interested in the paranormal? (tag: UFO) We have books on that."

There are some entertaining countermeasures that can be deployed to befuddle the bots trying to analyze your images. I often tag delicious items on my dinner plate with the names of people I know, and Facebook plays along, at least for now. However, soon it will probably ask "don't you mean Market East Coast Oysters on the Half Shell?" and offer to have more sent to me via an online food delivery company.

Even though most computer users understand at least some of the implications of posting and tagging photos, they seem to have decided, either consciously or implicitly, to go along with the game. Privacy experts disparagingly called folks like this "sheeple." They are also the ones who give up their email addresses for a free magazine subscription or answer a detailed online survey, hoping to win a $500 gift card. They make a Faustian bargain with online companies, allowing total access to their lives in exchange for services that appear to be completely free.

With improving technology, once even a single good photo of you is tagged with your real name on Facebook, your privacy is a goner. You will be identifiable and trackable—unless you are prepared for a face transplant, or at least to radically modify your appearance.

Realface Glamoflage shirts were designed to do exactly that. Artist Simone C. Niquille has created multi-face designs to distract and confuse facial recognition software. Her shirts have a number of faces, including that of Michael Jackson and Barack Obama, and, for now at least, seem to confuse the face bots.

An even more radical approach to dodging facial recognition cameras has been suggested by artist and researcher Zach Blas, with an idea he calls Facial Weaponization. He helps people create masks with weirdly-morphed versions of their actual face, hoping to bedevil the recognition software.

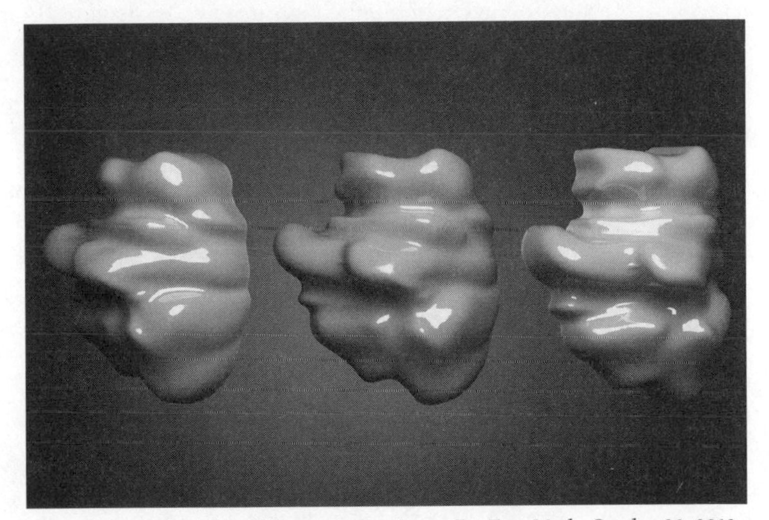

Figure 6. Zach Blas: Facial Weaponization Suite: *Fag Face Mask–October 20, 2012*, Los Angeles, CA. Courtesy of Christopher O'Leary.

Wearing masks in public can get you in trouble with the law in some places, but surely they cannot ban outlandish hair styles and creative makeup. Artist Adam Harvey has been experimenting with eye-catching patterns that put the facial recognition programs off your digital scent. He got his inspiration from Dazzle, a camouflage paint used on battleships in World War I.[103] In the same spirit, photographer Petr Prokop has created Face Dazzler, a smartphone app that distorts your face in photos. He claims they become immune to facial recognition programs, but are still recognizable to your friends and family.[104]

There are over seventy synonyms for "friend" in *Roget's Thesaurus*—"acquaintance, neighbor, well-wisher, advocate"—but none of them is "a person or entity you don't really know but who seems to share your prejudices so you clicked yes on their friend request." All technologies, and especially social media ones, expropriate familiar words and create their own creepy vocabularies. What is a poke? A news feed? A timeline? A like? Surely not the same as in the real world.

In a fascinating experiment, Dean Terry and Bradley Griffith at the University of Texas, Dallas created the concept of a Facebook enemy. Using their EnemyGraph application, users can declare their undying hatred of a politician, musician, or habit like smoking and see how many others chose the same enemy. Justin Bieber makes the list twice because he has multiple incarnations on Facebook.

For a while, whimsical entries such as "truck balls" and "bunnies" made the "Top Enemies" list, but as the EnemyGraph app attracted more users, the true enemies of the people rose to the top. Facebook normally bans applications that could disrupt the monetization of their ever-growing network of connected human beings. So if anyone tries to launch an "unfriend everybody" functionality it is quickly shot down by the company.

Yet Facebook tolerated EnemyGraph, perhaps because someone there understood that it was unwrapping a whole new category of information. Users could now be grouped by their mutual dislike of Bieber or bunnies or Microsoft, so targeted ads for things like bunny eradication powder might be both possible and probable.

Terry noted on his website that "people are also connected and motivated by things they dislike. Alliances are created, conversations are generated, friendships are stressed, stretched, and/or enhanced."[105]

While not ignoring the commercial possibilities of the data he is generating, he says it is really just a fun social experiment and critique of the philosophy of Facebook: "So, Facebook runs queries to find affinities. EnemyGraph runs what we call dissonance queries. So if you have said you like, say, *Portlandia* (a TV show) on your profile page, and in our app one of your friends has declared them an 'enemy,' we will post this 'dissonance report' in the app. In other words we point out a difference you have with a friend and offer it up for conversation, as opposed to a similarity. Relationships always include differences, and often these differences are a critical part of the fabric of a friendship." He goes on to suggest that in the "country club atmosphere" of Facebook these differences are ignored because dissonance is not part of their "social philosophy."

Some are even suggesting that the digital trail you leave simply by "Liking" things on Facebook can paint a fairly accurate profile of you. The website www.youarewhatyoulike.com tells you about yourself based on your Facebook likes. I dug into the logic of the program using a specially constructed new Facebook profile to see which likes it used to form its opinion. It reported two of the ten Likes on the profile, and they were both travel sites. It told me pleasant things like I am "warm and trusting" and "liberal and artistic."

The algorithm behind this comes from a research project at the University of Cambridge. For marketers, this kind of application is real gold. You can be sure if this level of analytics is available to all of us, for free, what they have is a lot more powerful.

While this application is fun and basically harmless, you may not want your current or a potential employer poking around on your Facebook profile, Twitter feed, personal blog, or photo albums. In 2013, BuzzFeed and CNN collaborated on an article called "ten people who learned social media can get you fired."[106] Examples included "The Bitter Barista" who blogged what he really thought about his customers. Former California Pizza Kitchen server "Timothy DeLaGhetto" tweeted as @Traphik about how little he liked the uniform he was required to wear. He no longer has to wear it since he no longer works for the company.

I once found myself at the very center of a photo mystery that was bedeviling tech journalists. "Who is this geek supermodel?" they were asking as the same woman's face appeared in both Microsoft ads and promotional materials for archrival LinuxWorld. People were wondering why Microsoft would use a model that was also being used by other tech companies.

"Nobody knows her name," Robert MacMillan wrote on wired.com. "With immaculately coiffed blond hair, striking black glasses and a perky grin, she's an idealized image of the geek girl next door."[107] One industrial psychologist said she was beautiful but also somehow approachable, the kind of girl that a guy who has been coding in his parents' basement for three days could still fantasize about asking out.

I was able to solve the mystery because she was sitting right in front of me in a University of Calgary classroom. Her name is Marla, and, in addition to being a student, she had a job at a local stock photo images firm. "One day," she told me, "they came around with a camera and took pictures of all of us and we signed model releases."

Her employer was taken over by Getty Images, which continued to market Marla's photo. A raft of tech firms decided to use her image in their advertising, often choosing the cheaper "royalty free" option, which meant other firms could use it too.

"It's become kind of a 'Where's Waldo' type of thing," Marla told me, "with friends emailing me to tell me about places where they've seen my image." I had a lot of fun breaking this story, and enjoyed her comment that "I love life when it throws bizarre incidents like this my way … I fully recognize that this is my 15 minutes of pathetic fame—so I'm savoring every moment!"[108]

Revisiting the Marla story a decade later reveals the disturbing way in which photos can persist. The websites for LinuxWorld Expo 2003 and Microsoft's brochure from that era are long gone. But Getty Images still has her image for sale, and why wouldn't they? She was a best-seller, even though she told me she never received any royalties for her photo.

For the "royalty free" license fee of less than ten dollars, you too can use Geek Supermodel Marla, at least the way she appeared in 2003, in your next advertising campaign:

Figure 7. Marla in her original form. Getty Images, photo E013748, used under license.

I decided to see if Marla, or at least her hard-working photo, was still on the job. Luckily, we now have tools for that very purpose. One is Google Images search, which allows you to plunk in an image and traverse the Internet, looking for uses of it. This comes in handy for companies like Getty Images to track the use of their photos online.

Doing that turned up seventeen hits, including a staffing agency in Lee's Summit, MO; a web developer in Saratoga, CA; and dentists in both American Fork, UT, and Santa Rosa Beach, FL. She does have lovely teeth but I wonder if the dentists have paid to use her smiling face?

Popping Marla's iconic photo into TinEye, a Toronto-based competitor to Google Images, produced other uses of her image, from as far away as Poland, Korea, and Japan, mostly related to a Kyocera Zio smartphones ad campaign in 2010. While almost all Microsoft sites have long since dropped Marla, Microsoft New Zealand is still using her smiling face to promote ancient versions of its products.[109]

We could find even more Marla ads if we used the Wayback Machine, whose mission is to save as much of the Internet's content as possible for posterity. Just go to www.archive.org, plug in a website, and select a date—you are able to travel through time, browsing cached web pages that have been preserved for eternity.

We used to be able to keep our pictures private. They were in a camera, a trusted photo lab, an album, or a dresser drawer. But even if we are scrupulous about not posting anything online, hackers and even automated technology can spread our photos far and wide. For instance, a Trojan Horse call PixSteal can sneak into your computer and send all the photos it finds to an FTP server waiting somewhere in the world. Coupled with your IP address, and soon, effective facial recognition, the bad guys behind PixSteal are in a perfect position to blackmail you if there is even a single picture on your computer that you would not want made public.

The motivation to find out more and more about you has become a matter of Dollars and Pounds and Euros, Yen, and Renminbi. If Professor Acquisti can de-anonymize people on dating sites from their Facebook profile shots, driven solely by academic curiosity and using publicly available resources, what can somebody do with your personal information when there is real money on the line? A photograph is, after all, an extension of our sense of sight over space and time. But we have other senses too.

# Sensor Creep

We have been extending our senses as far as technology will allow since Galileo and his contemporaries turned their optical instruments skyward. Telescopes and binoculars can be used for good (a sailor finding land) and evil (an unwanted voyeur). Sometimes the virtue or vice depends on your point of view. The hunter thinks his binoculars and telescopic gun sight are wonderful tools; the deer in the cross-hairs feels differently.

Everywhere we go, swarms of sensors are watching us. They are in the road, the signs, and the streetlamps. They are in your dishwasher and will soon be in your toothbrush. They are definitely in that "red light camera" that just snapped your car's picture. You know about that one because you saw a flash and are now awaiting the bad news in the mail. However, most sensors are silent, unlabeled, and often almost invisible. They are talking to each other all the time. Sometimes they let us in on the conversation, sometimes they do not.

Experts call this matrix "The Internet of Things," and it is a hot topic whenever techies get together. When Rob van Kranenburg, a member of the European Commission's IoT expert group, raises the possibility of "non-invasive neurosensors scanning your brain for over-activity in every street," most people get a decidedly creepy feeling. Yet we seem ready to accept radio frequency identification (RFID) chips in our passports, our clothing, and even in medical devices that go inside our bodies.[110]

Giving a unique Internet address to almost everything actually required changing the fundamental numbering system of the online world. Back in 1981, the designers of the Internet Protocol could not conceive of enough computers in the world to exceed the 4,294,967,296 Internet addresses they provided. But when you think about every car, toaster, streetlight, and school kid having one, we have pretty much run out.

The new system, called IPv6, theoretically accommodates a whopping 340,282,366,920,938,463,463,374,607,431,768,211,456 unique addresses, which should certainly allow all our garage doors, refrigerators, and toothbrushes to be connected and talk to each other. But should they?

Some sensors are clearly beneficial. The ones that alert you that your house is filling with smoke or carbon monoxide, for instance, are definitely your friends. The sensors that deploy the airbags in your car faster than human reaction time have saved countless lives. Yet those airbag sensors start to suggest the darker side of sensors.

If your car's airbag sensor alerts emergency services that you have had a serious crash, this may well save your life. But what if it also records the detailed state of your body at the time of the accident? What if it quietly takes a breath sample or forwards your most recent cell phone calls to the authorities?

Writing about the near future, CNN predicts sensors in automobiles will measure vital signs "such as heart rate, eye movements and brain activity to detect everything from sleepiness to a heart attack."[111] The article also notes that "Nissan is experimenting with an array of technology that detects drunken driving. A sensor in the transmission shift knob can measure the level of alcohol in a driver's sweat, while the car's navigation system can sound an alarm if it detects erratic driving, such as weaving across lanes."

One can easily imagine injured crash victims frantically pawing at the wires under the dashboard to abort certain revelatory transmissions. The scene gets even more unsettling if the sensors and their associated systems do their work behind the scenes, without your knowledge.

One of the creepiest features of the Internet of Things is how you may unwittingly become part of it yourself. Noting that GPS chips are now smaller than a match head, and keep getting cheaper, blogger

John Brownlee predicts that "we're fast zooming into a day and age where GPS nano-chips will be sprayable in a fine mist all over your body as you pass through airports customs."[112]

It is enough to make you want to skip airports altogether and just jump in your car. That will come with its own surveillance issues, even if you obey all the laws and do not have an accident. License plate readers are proliferating, and there is even serious talk of tracking toll road users with them. While fugitive pursuits and speeding tickets would be the obvious applications of this kind of technology, there are more subtle ones. Oregon, realizing that fuel-efficient cars use less gasoline per mile, and that electric cars use none at all, is fretting about how to equitably collect road taxes. One possibility: attaching a meter to the car's diagnostic system to track miles driven. Privacy advocates say that somebody would quickly decide to give drivers discounts for avoiding congested roads, creating a *de facto* GPS tracking system for drivers.[113]

We might well get to the point of coming home from a road trip to read our e-tickets, speeding fines, and road use charges. What about simply going for a walk? Authorities cannot tax our strolls yet, though they certainly might monitor them. Why would they do that? Perhaps to make sure we are getting the amount of daily exercise we promised our insurance company when we opted for the "active person" medical and life insurance policy. There are already plenty of apps that track your exercise, including some where you are fined if you do not meet your goals.

In reality, we would not want our sensors to tell us every time they take a reading or communicate with another system. Imagine if Google's self-driving car pestered you every time its sophisticated sensors scanned the road ahead. This new fleet of driverless vehicles has already logged over half a million miles virtually accident free. Once you accept that your life is in the hands of a bunch of experimental technology, riding in one is reportedly quite relaxing.

The day will come when our cars drop us off at our destinations and then scurry off to park themselves. Perhaps they will have car-to-car conversations in the parkade. They might even joke about what we humans were doing in the back seat. Most of us would be fine with one car telling another "I'm about to vacate stall #216." But what happens when your car brags about the maximum speed it has attained today, and a nearby police car is listening in to the vehicular banter?

Eric Gauthier was driving his new Pontiac Sunfire in downtown Montreal in April 2001 when his car struck another vehicle and killed its driver. With no witnesses, and a denial by Gauthier, police and the crown prosecutor sought to use data from the car's event data recorder (EDR). It showed the car was traveling between 130 and 160 kilometers per hour, well over the speed limit.

The EDR, which Gauthier probably did not even realize he owned, records key parameters, such as speed and whether or not the brakes are applied whenever an airbag is deployed. The original intent was for engineers to analyze the statistics, but now police and prosecutors want to see information from an EDR accepted in court. The EDR data was admitted in the Gauthier case and he was convicted of dangerous driving.[114]

These automotive "black boxes" have also been accepted as evidence in the U.S., U.K., and Australia. There is certainly pressure to have them routinely accepted as evidence, since they are objective and probably more accurate than human memory.

Many drivers are actually inviting Big Brother under the hood in hope of saving money on car insurance. They are installing "driving monitors" such as the "Snapshot" from the Mayfield Village, OH-based Progressive Corporation. These tell your insurance company certain things about your driving such as the distance driven, when you travel (midnight to 4 AM is bad; the roads are full of drunks), and instances of "hard braking" (which they define as a change of more than seven miles per hour in one second).

While slamming on the brakes may mark you as an aggressive driver, there can be good reasons, such as avoiding a child who runs into the street or a last-minute red light. A number of bloggers who have installed the Progressive Snapshot say it has elevated their stress level while driving. Others have pointed out that this is really "usage-based" insurance since a major factor is the distance you drive per year.

According to a speaker at the Telematics Update conference in Chicago, the public's concern about creepy government spying has led to privacy concerns about driving monitors. Joe Reifel, an AT Kearney partner, said that they are predicting an adoption rate of 22% over the next three years, down from a previous forecast of 30%.[115] Snapshot records data through the OBDII diagnostic port, standard equipment on most cars manufactured since 1996, and sends it wirelessly to the company. Because of the legal concerns, they do not use GPS tracking, which could yield much more interesting information about where you are driving (such as nightclubs, liquor stores, and racetracks).

Progressive maintains that the Snapshot is a "win/win" because clients may get a discount, but no matter what it reveals, the Snapshot data alone will not make their rates go up. Prospective clients who test it for a month are then given a rate quote and the option to keep using the device.

A voluntary technology like this can easily become "*de facto* mandatory." Since insurance spreads risk over a pool of drivers, declining to have the monitor installed may eventually lead companies to make assumptions about your driving habits. There is no doubt that new cars will have the ability to collect and send out data about your driving habits. The big question is—who will have access to it?

The first documented use of evidence from a Snapshot device in court was in a murder trial in Parma Heights, OH. Michael Beard was accused of suffocating his infant daughter. He remembered that he was on the thirty-day Progressive Snapshot trial, and the timestamps in it provided evidence (to the jury's satisfaction) that he was not in the house long enough to do the deed.

Ironically, "Beard's driving made him ineligible for any rewards," according to a published report on the case. However, when he remembered that he had forgotten to take out the Snapshot, "I knew if I could retrieve the information I could prove I wasn't there. Progressive told me that after so long they usually clear the information—but when they told me they still had it, 'Oh my God' was all I could say."[116]

Progressive is not the only insurance company getting on the car tracker bandwagon. In 2010, Allstate announced a similar device it calls Drivewise. Like Progressive, Allstate avoids using GPS to track drivers. But that might change. Allstate's CEO Tom Wilson has been quoted as saying "we're going to do something with teen drivers so you can actually know where your kids are if they're driving."[117]

There are already several apps that do just that. Most of them rely on the fact that the average teenager would rather leave the house naked than without a cell phone. Nervous moms and dads can, openly or secretly, install GPS tracking software on their children's phones. Life360, whose basic version is free, combines data from smartphones and car tracking devices, just like they use in the movies. Results are neatly displayed on a webpage or via the smartphone app itself.

Most rental cars have a hidden tracking device so the company can find their vehicle if some deadbeat leaves it abandoned in a ditch. This is a separate GPS from the one that they rent to you at the counter. You can, and probably should, clear out the data from that one before you return the car. The other one is hidden and not accessible to the renter.

According to a report in the trade publication *Auto Rental News*, some agencies even have a remote kill switch on their vehicles. "Al Llanes of Global Rental Car of South Florida Inc. restricts his renters to the state of Florida," the magazine reports.[118] "He uses his tracking system to set up a virtual perimeter (or 'geofence') that alerts him when the state line is crossed." What happens when the car goes over the line? The article says that Llanes remotely disables the vehicle, and usually receives a phone call from the renter. He offers to restart the vehicle, but points out that there will be an extra mileage charge.

On the other side of the U.S., the out-of-state mileage charge struck renter Ron Lee, who was presented with a bill for over $1,700 for what he thought was about a $150 rental. The difference was an obscure dollar a mile surcharge for taking the car out of the state of California. The vehicle's GPS ratted him out.[119]

Other rental car companies have tried adding "speeding surcharges" based on GPS data collected by the car. In a Connecticut case, an independent hearing officer estimated that the real cost of speeding in terms of extra wear and tear on the vehicle was thirty-seven cents, well below the $150 that the company was charging as a speeding penalty. Acme Rent-A-Car was ordered to stop fining speeders.[120]

Aside from privacy concerns, a bigger issue arises from what a rental car company might do with all that data it collects on its customers. Even in aggregated form, with no personal information attached, the data that rental companies collect can be very valuable. A log of precisely where thousands of rental cars have been driven, where they have stopped to admire the view, and where they buy gas would be very valuable to someone trying to choose a site for a roadside service area. Just as the ancillary revenue from running an airline reservation system can out-pace the profits from flying planes, selling data on customer behavior may become a major cash cow for car rental firms and other travel providers. With the right tools, someone might be able to "torture" the aggregated data to find out about a specific individual.

While working with employees of a mid-sized Canadian city, I asked them what disturbing things they see in their jobs. "City vehicles all have GPS sensors on them," one employee piped up. "The snowplow drivers are afraid they'll get in trouble if they take a break to warm up and grab a coffee." Given that this city endures major snowstorms and Arctic temperatures, you might think a trip to Tim Horton's is a reasonable request. The problem is that there is no clear policy. Drivers know the unflinching eye of the GPS system is always on them, and they fear it.

Public servants trying to solve real problems increasingly find themselves relying on technology that has the potential for serious abuse. A well-intentioned plan by British Columbia transit operator TransLink to use cell-phone "pings" from drivers' phones to improve its real-time traffic maps raised privacy hackles. Even though the transit operator swore that nobody could be identified, people howled in protest.[121]

Intellistreets street lamps, manufactured by Illuminating Concepts of Farmington Hills, MI, dispense a lot more than illumination. They are actually sensor-enabled two-way communications devices. They can broadcast music, and even tell buildings to dim their lights when there are no people around. Security guards can talk through their "Concealed Placement Speakers" as disembodied voices.

The manufacturer's home page also offers "a wide range of sensors (that) can be utilized for exciting pedestrian user interaction." Accessories for your new streetlamp include CBRNE (chemical, biological, radiological, nuclear, and explosives) detectors as well as a "shared data strategy (that) analyzes images between sensors and can direct PTZ (pan/tilt/zoom) cameras and situational awareness to the end user." If your staff or other persons of interest (for example, prisoners) are sporting RFID tags, this system can also identify and track them for you.

I witnessed firsthand just how exciting, if only marginally effective, a talking security system can be while walking through a public housing project in Washington, D.C. Probably unable to afford those $3,000 streetlamps, the complex was using old-fashioned security cameras with nearby wall-mounted loudspeakers, all connected to an unseen guard room. "Hey, you in the red hoodie—no skateboarding here." A shocked look from the kid was followed by ... more skateboarding. I did not stick around to see if live security action followed.

The U.K. has also dabbled in talking surveillance cameras, even holding "competitions for children to become the voice of the camera."[122] Presumably having your grandkid tell you to pick up after your dog will have a stronger impact than the lady who tells you to "Mind the Gap" or the man who urges you to "Please Stand Clear of the Closing Doors."

I give a lot of talks to school groups, and often learn amazing things from them. One eager sixth grader asked me, "Did you know that there are cameras in the eyes of the mannequins at The Bay?" referring to Canada's iconic department store. I said I didn't think that was true and he quickly replied, "Yes, there are! That's how my sister got busted for shoplifting." The teacher stepped in to end this oversharing of family secrets.

If we allow what technology makes possible, your nearest streetlight or trash bin may have the same capabilities envisioned by English philosopher Jeremy Bentham for his Panopticon, the perfectly designed prison in which jailers manage their charges by the simple stratagem of being all-seeing while remaining unseen themselves.

Yet, it would be oversimplifying to suggest that we are moving to a world of government wardens and citizen prisoners. After all, many of the security measures that are put in place are put there at the request of citizens like us to attempt to deter crime, terrorism, and other bad behavior.

I once got to chat with Anthony Zuiker, creator and executive producer of the *CSI: Crime Scene Investigation* television dramas. He assured me that "everything we do on the show is based in science, but sometimes we do speed it up for television." He also said he is well aware of the "CSI effect"—the rising expectation that high tech forensics will be available and applied even in minor cases. People who have a hundred-dollar GPS unit stolen from their car are thus outraged when police don't dust for fingerprints or search for traces of the DNA left by the thief.

Our growing reliance on "always-on surveillance" is illustrated by the tragic case of teenager Kendrick Johnson, who was found dead, wrapped in a gym mat, at his school in Georgia. His parents demanded the video from the school's surveillance cameras, even launching a Twitter campaign with the hashtag #giveusthetapes when they were not immediately provided.

When police released the evidence, there were gaps that some said were due to the fact that the cameras were motion activated. However, the victim's parents believed that there was tampering. CNN took the tapes to Seattle-based Grant Fredericks, a certified forensic video analyst who also played a role in the analysis of the Vancouver 2011 riot tapes. Fredericks found that the videos did not appear to have been edited, but he also questioned the quality of the data since it was not provided in its original form.[123] We not only expect that surveillance footage will be there to help us solve crimes—we get angry when it does not show us what we want to see.

Cameras and sensors in smartphones have already been pressed into service as medical devices, testing eyesight (Vision Test 3.0 was the iPhone "medical app of 2010"), blood pressure, glucose levels, and even heart rhythms. However, no smartphone sensors can approach the sheer existential weirdness of the "smart toilets" sold mainly in Japan. One, from Toto, weighs you when you sit down, checks your body temperature, and does on-the-spot urinalysis. Some of its throne-like competitors will also send all of this information electronically to your computer or directly to your doctor's office.

According to one report, Kyushu-based Toto, which introduced its first smart toilet in 2005, has racked up 10,000 units of sales, even though the toilets can cost $5,000 or more. And, yes, they can be hacked. It is done by Bluetooth. The MySatis Android app was designed to give you full control of your pricey Satis commode, even allowing you to play the perfect music during your experience and keep a "toilet diary." A new report explains that "the advanced 'Satis' automatic toilet can be remotely operated by a free app available on

Android smartphones that lets pranksters raise and lower the toilet set as well as trigger a bidet function and flush."[124]

If you live long enough, you may well wind up with sensors in your underwear, or your adult diapers. A Spanish company SiempreSecos, which translates as "always dry," has created a urine sensor with a companion wristband to alert you, or your caregiver, that it is time for a change. Let's just hope you're not sitting down to a formal dinner or something when the thing goes off.[125]

There is definitely an upside to sensor technology. David Webster, a partner in the California-based design firm IDEO, waxes poetic about "being able to capture and track and pattern-recognize body data, which can be measured by devices in and on the body, and activated as data streams in the cloud."[126]

Webster predicts that sensor technology will lead to a new era of "precision medicine" with "the potential to change everything in a really positive way." Having a sensor in your body that detects when you are sick may be great, if it alerts you and you take appropriate action. If it sends clandestine messages to your insurance company, or even your favorite pizza parlor, however, then it crosses the line from useful to insidious.

In the chilling YouTube video "Ordering Pizza in the Future," a fictitious caller tries to order a Double Meat Pizza, but the order taker knows far too much about him. He is told "there will be a new $20 charge" because "your medical records indicate you have high blood pressure and extremely high cholesterol. Luckily we have a new agreement with your national health care provider that allows us to sell you double meat pies as long as you agree to waive all future claims of liability."[127]

The pizza girl also knows his waist size, his recent library checkouts, and that his credit cards are maxed out. As a final blow, he is informed that there will be a danger zone charge due to the fact that he now lives in a high crime zone due a recent robbery.

The ease and cheapness of connecting everything to everything means there is no technical reason why anything digital in your life cannot be connected to anything else, with or without your knowledge.

Over a decade ago, appliance makers were eagerly anticipating the day when the refrigerator would tell the stove what it has on hand ("beer, expired milk, baking soda") and then they would decide what you could make for dinner ("nothing worth eating"). Perhaps they might even send you a text suggesting a visit to the supermarket, or, for that matter, place the grocery order for you.

Consumers found the idea of a smart kitchen interesting, but not worth the extra money. The promotional videos for the Samsung T900 and LG Smart Thinq™ refrigerators provide a somewhat creepy demonstration of how engineers imagine people might use refrigerators, but with little connection to real family life.

LG's $3,500 monster fridge boasts an 8" LCD screen where you can laboriously log the products you intend to buy, or have bought, along with their expiration dates. Or you can simply toss the milk in there and smell it once in a while. It seems that people will incorporate new technologies into their lives if they see clear and desirable benefits that outweigh any downside. Smartphones, GPS units, and applications like Google Maps clearly fit into the "benefits outweigh the negatives" category.

Sometimes, however, the risks of sensor technology may be hard to discern. Especially when the devices are going on, or even inside, our bodies.

I had the opportunity to advise the organizers of the 1988 Winter Olympic Games about ways to make their television coverage really unique. "You could put cameras on the athletes and sensors inside their bodies," I suggested. "Not only could we see the event from their viewpoint, the sensors could tell the audience in real time about their pulse, blood pressure, and other biological parameters."

"OK, as long as they stay above the waist," laughed an Olympic gold medalist swimmer, who was there representing the athletes.

Today, it is trivially easy to monitor all kinds of physiological parameters of athletes, or anyone else.

Nike's FuelBand exercise monitor wristband uses oxygen kinetics to measure activity and can even track minute movements during sleep. It connects to your smartphone to upload its data. Gizmodo writer Adam Clarke Estes caught a lot of people's attention with a piece titled "Your FuelBand Knows When You're Having Sex," even suggesting "they could also tell when you fake an orgasm and possibly alert your spouse or significant other if you're having an affair."[128]

Writing on TechCrunch, Gregory Ferenstein muses along the same lines: "were I married, my wife might like to know why I burned 100 calories between 1:07 to 2:00 am, without taking a single step, and fell asleep right afterwards."[129]

You *could* take the device off before leaping into the sack. But that would produce a telltale and potentially incriminating "data gap." As we start to accept devices that are intended to be "always on," turning them off starts to raise questions.

Seeing an opportunity to lead the world, after being racked by successive earthquakes, Christchurch, New Zealand is planning to spend billions to rebuild itself, and, in the process, to become one of the world's first "Sensing Cities." Getting an early warning of the next big earthquake is clearly very high on the priority list for Christchurch; but they intend to go far beyond seismic sensors.

Although they claim they do not intend to track individuals, the project's webpage has the creepy suggestion that "Arguably the technology to track an individual is already widely accepted by society, with cellphones carried by the majority of the population."[130]

Sensors are creepy in many ways. They are surreptitious, often invisible, persistent, always on, and always monitoring. Often, they work for the highest bidder, and may show no allegiance to their owners.

There certainly is potential for objects in our lives to interact in helpful ways. A smart scale, or a toilet that does an instant analysis of your bodily fluids, could tell the toaster "no more bagels" if you are trying to reach a diet goal. The question is—do you want it to? What has been missing so far is a kind of meta-language and social norms to effectively tell our electronic servants which features we think are cool, and which are creepy.

Increasingly, though, you will have no choice—the surveillance will be automatic, continuous, silent, pervasive, and non-consensual. Even if you only want to buy a loaf of bread or some razor blades, as people in a Tesco store discovered.

# Tracking Creep

For a short while in 2003, the possible future of retail shopping was revealed in all its creepy glory in a Cambridge, U.K. food store. Supermarket giant Tesco set up a clandestine system that, according to a newspaper account, "triggered a CCTV camera when a packet of Mach3 blades was removed from the shelf. A second camera took a picture at the checkout. Security staff then compared the two images, raising the possibility they could be used to prevent theft."[131]

The company claimed they were just trying to improve customer service by keeping track of inventory, but when the store's manager provided photos of a shoplifter to authorities, privacy advocates condemned the system. There were protests outside the store and in the media. Gillette, the manufacturer of the blades, was also taken to task for facilitating the scheme by putting RFID tags in individual product packages.

Now, a decade later, if you buy a book or an electronic gadget, you will probably see a mysterious small paper square with an embedded antenna fall out of the packaging. RFID chips are even sewn into designer clothing. High-end fashion retailer Burberry has implemented RFID tagging to "enhance the customer experience in selected stores."[132] Their website also explains that if you do not want to walk around with a tracking tag in your clothing, you can deactivate it by "simply removing the textile RFID label"—as if that's the most natural thing to do to an $1,895 cashmere sweater you just bought.

A report in *The Economist* describes Burberry's "Customer 360 plan" which will allow the company to record the "buying history, shopping preferences and fashion phobias" of customers "in a digital profile, which can be accessed by sales staff using hand-held tablets."[133] The article notes that the database might prove to be a source of embarrassment, "for example if a customer who has bought racy gifts for his mistress enters a Burberry store with his wife and is enthusiastically ushered to the skimpy bikinis."

Mondelez International, whose "brand family" includes Chips Ahoy and Oreo cookies, is planning a 2015 launch of "smart shelves" which, according to the *Washington Post*, will work out your gender and approximate age in order to sell you products calculated to appeal to your demographic group.[134]

The shelf sensors will not take your picture, but that does not mean your privacy will be protected. By the time these devices arrive at your grocery store, the chances are good that you may have been convinced to carry a personal RFID chip, or link up your smartphone, to get a discount or loyalty club points. You are liable to be under electronic surveillance from the moment you enter the store. The retailer will be able to gather information on where you linger, what you pick up, what you put back, and of course, what you ultimately buy.

Two shopping malls, Promenade Temecula in California and Short Pump Town Center in Richmond, VA, recently announced plans to install a FootPath shopper tracking system to keep tabs on shoppers as they move around the stores. The technology, from the U.K. company Path Intelligence, is a way for brick-and-mortar retailers to keep up with online retailers, who can already track every click a customer, or even a casual window-shopper, makes in cyberspace.

Using customers' own cell phones, Path Intelligence tracks movement patterns of shoppers and, according to their webpage, "can accurately locate how those devices are moving around the store or mall." Their data is used to evaluate things like whether or not large anchor tenant stores such as Sears, who usually get a rental discount in shopping centers, actually attract customers to the smaller stores in the mall.

Path Intelligence has found, for example, that less than 40% of the people who visit an Apple retail store subsequently proceed to another mall tenant. Instead, most appear to simply take their new gadgets home and start buying stuff online. Critics, including U.S. Senator Charles E. Schumer, expressed concerns about shopper privacy. As a result, plans to implement Path Intelligence in those malls were shelved, at least temporarily.[135]

An even creepier test was carried out in London, U.K., in June 2013. Marketing firm Renew London planted sensors in trash cans that tracked the unique signature of every smartphone that passed by. According to a chirpy press release from the company, their Renew ORB system "provided a concise breakdown (to the 50th of a second) of the movement, type, direction, and speed of unique devices that the Renew Network gather across Renew ORB test sites, and help identify peak footfall times from key hotspots in the City of London."[136] The number they are capturing, the phone's MAC (Media Access Control) address, is a fingerprint that is specific to that device. In just one week, they grabbed over four million MAC addresses, over 500,000 of them being unique.

The company insisted in a statement that the addresses it collected are "anonymized and aggregated" but that is a bit misleading.[137] If the phone is activated on a network, the telecommunications provider has to know information about the device and usually about the owner.

As we are starting to realize from disclosures of Edward Snowden and other sources, information at the telecom company doesn't always stay at the telecom company. In addition, by simply tracking a phone, deductions can be made about the person carrying it. Visiting the ladies' or men's room provides your gender. Entering a bank branch may disclose where you have your accounts. What do twice weekly visits to a sexual diseases clinic signify?

Further evidence of the real commercial purpose of Renew ORB comes from a creepy video the company has posted.[138] The video's narrator explains that, in the online world, "cookies tell marketing teams about the kind of things that Jack (the animated person in the video) finds interesting." The cheery voice goes on to lament that "Cookies don't exist in the offline world—until now." By tracking Jack's movements as he innocently walks around, marketers in the physical world can claim some of the advantages their online competitors already have.

Another important piece of data that your smartphone is leaking is the list of every network it has been connected to. The phone keeps reaching out, looking to see if that network is in range, until you specifically clear it from its list.

Speaking at Black Hat Asia 2014 in Singapore, Glenn Wilkinson of SensePost's U.K. office showed how this information could be used to make inferences about the phone's owner. "If I see somebody whose phone is looking for the British Airways First Class Lounge network, I know he's probably a high roller," he said. "If his phone is seeking the employee network of the RBS (Royal Bank of Scotland), I have a pretty good idea about where he works. If his phone is also seeking the WiFi network at Hooters, well that's a problem."[139] Putting all this information together is getting easier and easier, and no laws are being broken by plucking information out of the public airwaves.

Wilkinson warned the audience that "we are all carrying around the most perfect surveillance device ever invented, completely voluntarily, right in our pockets." To dramatize the vulnerabilities, he deployed "Snoopy," a drone helicopter that can fly around taking pictures and intercepting wireless signals.[140] Using data collected at the conference, he was able to tell an audience member what part of Singapore he lived in and even show him a Google Maps photo of his street.

While it crosses onto the wrong side of the law, this technology can also be used to hijack wireless connections, and even insert false data. Wilkinson explained how it was possible to attack someone using a coffee shop's Wi-Fi. "You could replace every image on someone's smartphone with a picture of a cat," he suggested, "or continually flip their images upside down so you'd see them frantically turning their phones over and over."

Loose-lipped smartphones certainly give retailers new ways to track customers. However, many shoppers will probably be delighted to pin identification tags on themselves, in return for discounts or loyalty points. Walmart, which pioneered the use of RFID tags for inventory tracking, has been putting them on individual socks and underwear, supposedly so store staff can quickly replenish missing sizes. It is only logical that retailers would like to put chips on customers too.

In an alarming book called *Spychips: How Major Corporations and Government Plan to Track Your Every Purchase and Watch Your Every Move,* Katherine Albrecht and Liz McIntyre point out that even after you toss RFID product tags in the trash they can still be analyzed.

Albrecht also suggested in a radio interview that stores with RFID readers might also be able to read the chip in your passport or your enhanced driver's license, though there is no evidence retailers are actually doing that.[141] She worries that shoppers will be tracked "like rats in a maze" and she even suggests that as you approach a store display, the prices might change based on the neighborhood where you reside.

Many of Albrecht's speculations have already become reality. The *New York Times* reported that in the fall of 2012, the Seattle-based retailer Nordstrom, Inc. "started testing new technology that allowed it to track customers' movements by following the Wi-Fi signals from their smartphones." The company posted signs about this in its stores and customers quickly raised privacy concerns. While consumers may be made aware of retail tracking data, usually they have no control over how it is collected and used, aside from avoiding certain stores completely.

Lending even greater credence to Albrecht's speculation is a United States patent filed in 2006 and assigned to IBM Corporation.[142] The patent lays out a chilling near-future shopping scenario: "The RFID tag information collected from the person is correlated with transaction records stored in the transaction database according to known correlation algorithms. Based on the results of the correlation, the exact identity of the person or certain characteristics about the person can be determined. This information is used to monitor the movement of the person through the store or other areas."

There are more creepy U.S. patents such as #8,350,708 (Nike) which will allow your clothing to tell a treadmill that you're authorized to use it, and #7,925,549 (Accenture Global Services) which envisions "personalized advertising messages" routed to you based on, among other things, the RFID chips you happen to be carrying.

Perhaps the strangest place to find an RFID chip is inside medical products such as silicone breast implants. The Florida-based company Establishment Laboratories has introduced Motiva Implant Matrix Ergonomix which will allow suitably equipped doctors to read the name, rank, and serial number of the breast implants, in vivo.[143]

RFID technology gets even more invasive, and creepy, when combined with video cameras and sensors. Using these technologies, merchants can sort customers into gender and approximate age group, and track what they have picked up and put down. A Saint Petersburg, Russia-based company called Synqera even offers facial recognition and mood detection capabilities on its in-store terminals.

Intriguingly, one of the companies interested in working with stores to develop "customer tracking solutions" is Omnilink Systems, Inc. If you are a law-abiding citizen, you have probably never heard of them, nor their flagship OM 400 product, a bulky ankle bracelet. It is used to monitor criminals under house arrest and track those accused of crime who are out in the community. It even features a "mobile exclusion zone feature" where "you can not only track the whereabouts of convicted gang offenders individually, but also track the intersection of gang members restricted from association."[144]

A flurry of news stories in 2005 suggested that there would soon be a big market for RFID tagged baby clothing.[145] However, the idea never caught on, and has been basically superseded by the ability to track people with smartphone apps such as Life360. Not only will that one alert you to their whereabouts, it will even provide a warning when your children are approaching a known crime scene.

Technology has outstripped human bodies in countless ways. Cars move us faster than we can possibly run. Computers do complex arithmetic computations fast and flawlessly. IBM's Deep Blue defeats world chess champion Gary Kasparov who, in a typically human reaction, accuses it of cheating.

Vernor Vinge, Raymond Kurzweil, and other thinkers tell us that we are rapidly approaching "The Singularity," a point in time at which artificial intelligence will surpass human brainpower in important ways.

It is not obvious that we will even know when the machines have outstripped us. We also have no idea how they will treat us. Some science fiction writers predict that humans will be treated as a virus to be eradicated; others suggest that we will make excellent pets for our digital masters.

Most of us have already reached a kind of personal "privacy singularity." In some very important ways, our technologies, taken together, know more about us than our most significant human friend or lover. This is totally understandable, since we spend so much time exchanging information through technology. They are our electronic confidants, our faithful servants, and in some cases, two-timing spies.

Every time you "Look Inside" a book, "Like" a Facebook post, "Friend" someone new, send an email, or log on from a different place, you are leaving a digital trail that is being scrutinized to learn more about you. Techniques with exotic names like predictive analytics, k-means clustering, and cross-platform tracking provide deep insights into our thoughts and behavior. While human relationships may come and go, your online presence is forever.

For instance, let us compare the world of online shopping with going to a brick-and-mortar store. If you shop on a site like Amazon, you will undoubtedly be reminded about items that you looked at but did not purchase.

That's fair game. After all, a good salesperson in an upscale clothing store would also remember you and show you that tie you passed on last week, as well as others like it. The salesclerk would let you in on special sales and perhaps offer you some sort of incentive to buy it. So does Amazon. So far, the Internet is just mimicking real world commerce.

But what if that store clerk compared notes with the other retailers in a shopping mall to build a detailed dossier on you? What if conversations you had while walking between stores were mined for intelligence about your buying habits? A growing network of data partnerships is making this kind of snooping the norm in the online world. You probably agreed to it when you clicked on some Terms and Conditions that you did not really read.

As one blogger reported, "an image of some headphones I looked at (on) an e-commerce site ended up staring back at me from an ad on Facebook later that day."[146] Dave Obasanjo goes on to explain that was made possible by a system called Facebook Exchange (FBX). This is a real-time bidding platform where companies, like headphone maker SkullCandy, purchase access to your eyeballs through sponsored ads on Facebook.

Google's Gmail serves up ads targeted to your interests, which it can determine by having programs scan the content of your emails for certain keywords. Google notes that you gave them permission to do this when you signed up for Gmail. But what if a non-Gmail user sends an email to somebody on Gmail? The sender never agreed to those terms of service. A lawsuit against Google suggested this was a form of wiretapping. In reply, the company's lawyers asserted that they have that right to scan those emails anyway.

Google's Motion to Dismiss stated that "Just as a sender of a letter to a business colleague cannot be surprised that the recipient's assistant opens the letter, people who use web-based email today cannot be surprised if their communications are processed by the recipient's ECS [electronic communication system] provider in the course of delivery."[147]

Still, this does feel like a human salesperson trading on private conversations you had in the hall, or digging through your garbage. Even if the judges rule that the law is on Google's side here, that may just mean that the laws do not reflect our common sense understanding of what privacy is and how it should be protected.

The other major force in automating the online sales process is to use every scrap of information about your friends and their likes and dislikes. Amazon now alerts you if any of your Facebook connections have written reviews of products you are looking at. They also show you what "Customers Who Bought Items in Your Recent History Also Bought."

This is their "item-to-item collaborative filtering" algorithm at work. It attempts to figure out what you would like to buy based on the behavior of other people who are somehow like you. The success of this technique depends on having a good way to define "people like me" and a large and rich data source upon which to base recommendations. According to its April 13, 2012 Privacy Policy, Amazon collects the following automatically from each web visitor: "The Internet protocol (IP) address used to connect your computer to the Internet; login; e-mail address; password; computer and connection information such as browser type, version, and time zone setting, browser plug-in types and versions, operating system, and platform; purchase history, which we sometimes aggregate with similar information from other customers to create features like Top Sellers; the full Uniform Resource Locator (URL) clickstream to, through, and from our Web site, including date and time; cookie number; products you viewed or searched for; and the phone number you used to call our 800 number."

Amazon also collects your "length of visits to certain pages" which does have a certain "looking over my shoulder" feel to it. Then there is all the information you provided voluntarily. Did you write a review of a product? Put something on a Wish List? Ask to be reminded of a special occasion? It's all in your personal file at Amazon. The divergence from the friendly necktie salesman becomes even greater when you realize that this file is for keeps (unless you change your email address and even then you might be linked by IP address or credit card).

The computers at Amazon and Google usually have no reason to forget or purge any data, and every reason to hang onto it in perpetuity, a license you explicitly granted them when you agreed to their terms and conditions. You also agreed to have your data shared, sold, aggregated with the data of others, and sold again.

The power of social data is illustrated by the fact that some hotel chains employ full-time social media specialists to help manage their online reputations, and to respond to negative reviews on sites like Tripadvisor with comments like "yes we had some bed bugs but we've changed all the mattresses."

I use an itinerary management site called Worldmate, and always steer clear of the "Share with your friends on Facebook" button. I don't necessarily want a bunch of people I barely know having my detailed itinerary.

Over at Google-owned YouTube, it has been reported by insiders that "recommendations account for about 60% of all video clicks from the home page."[148] The Google employees who wrote this paper also note that "many of the interesting videos on YouTube have a short life cycle going from upload to viral in the order of days requiring constant freshness of recommendation."

If you want to see how much data is being collected on you, and how you are being tracked, just install a free application called Ghostery. It discloses, for example, that even the staid Government of Canada website is feeding information to Google Analytics, in return for what they get back—information on visitors to their webpages. And www.cnn.com is positively bleeding information about you all over the place. Accessed from New York City, Ghostery reported forty-three trackers on that site, some with revealing names like Visual Revenue and Facebook Beacon, but also some cryptic ones like Moat and Rocket Fuel.

The documentary film *Terms and Conditions May Apply* includes an estimate of the annual value of the data you give up to Google by using their "free" services. How the film's producers came up with the number $500 is as shrouded in mystery as the internal workings of Google, but there is little doubt that we are becoming more valuable every year as advertisers move from the traditional "show it to everybody and hope someone buys" model to highly targeted ads.

An astounding number of the creepiest technologies come from Japan, where they are accepted with a remarkable *sangfroid*, or perhaps just naughty delight. This is, after all, the country that had to invoke its "Antique Dealings Law" in 1993 to curb the sale of used schoolgirl panties in automated machines.[149] Cognoscenti say you can still find those machines in rural Japan, and pictures of them are preserved for posterity in YouTube videos.[150] Then again, with e-commerce sites like myusedpantystore.com, why would anyone even bother looking for a vending machine?

What could be technocreepier than buying schoolgirl panties from a vending machine on the way to work? Perhaps having your boss catch you sniffing them in your cubicle.

Tokyo-based KDDI Corporation has crafted an exquisitely creepy piece of software that might allow just that. It uses an employee's smartphone to figure out what the person is doing on company time, on a second-by-second basis. According to Addy Dugdale's piece in *Fast Company*, the "software is embedded into an employee's mobile that is connected to a server that analyzes their movements via the phone's accelerometer. At first, workers will have to input just what action they are performing into their mobiles so that their movements can be interpreted. The system becomes more accurate as time goes on, recognizing each individual's movements."[151] Employers do have many legal rights with respect to pieces of technology they supply to employees, but checking how your phone is jiggling or not jiggling certainly borders on the totalitarian.

I knew an engineer who could cobble together just about anything electrical or electronic. His front lawn had an elaborate system of sensors to detect intruders. "I didn't want to hit them with missiles or lasers or anything," he told me, "so if somebody stepped on my lawn the sprinkler in that zone would turn on. They knew that I knew they were there, and that was enough."

He also figured out how to detect, based on body size, which family member had just entered a room. Not only did his contraption flick on the lights, it turned on the TV, thoughtfully tuning it to the person's favorite channel.

"Get rid of that or I'm getting a divorce," his wife told him. "It creeps me out."

In a similar fashion, bars such as the Beantown Pub in Boston have been equipped with a suite of cameras that do "facial detection."[152] The goal is not to figure out who you are, but to figure out how busy the place is, as well as the ratio of males to females currently in the establishment and their approximate ages. It would be a small step to add a deeper analysis of physical characteristics, and undoubtedly some entrepreneur is working on just such a system.

Passwords, with all their shortcomings, do a reasonable job of protecting our workplace systems and even let us do online shopping and banking. But if Motorola's Research Division (now owned by Google) has its way, you may take a "password pill" every day. Designed to pass through your digestive system, it will emit radio waves and relieve of us of the chore of generating and remembering good passwords, as well as the guilt if we use "123456" or "snoopy."

The pill has been hailed as a great solution to the proliferation of passwords and damned as a frivolous invasion of your body. Made by Proteus Digital Health of Redwood City, CA, the digestible chip might find application in prisons and nursing homes, but it is hard to imagine a free citizen wanting to pop one of these with their daily vitamins. Aside from the unsavory feeling of having a foreign object broadcasting from inside your tummy, this concept might reduce stealing your access credentials to the simple act of grabbing one of your pills.

Matthew Drake of the Atlanta-based advertising agency 22squared thinks he has an even better idea. Speaking at TechCrunch SF 2013, he started his demo with "I've got a real problem. I love spending money but I hate wearing pants. More specifically I hate carrying around a phone and a wallet."[153]

Drake went on to demonstrate how he thinks we will be paying for purchases in the future. He predicts stores will install inexpensive Leap Motion controllers which will allow you to pay for purchases by waving, reaching, and grabbing imaginary objects in the air. He suggests each of us will come up with a "secret handshake" which will then serve as our password.

I predict that far too many people will choose gestures that involve the middle finger and will be using it in anger when the system does not quite recognize their finger dance. And they just might get into trouble if they use that finger sequence in public.

A sophisticated tracking system under development at the University of Nevada uses the Microsoft Kinect consumer product and "turns the user's hands into versatile sensing rods."[154] GIST (Gestural Interface for Remote Spatial Perception) can already sense colors and the proximity of human beings. In a video demonstration, GIST helps a blind person find her water bottle by giving her voice commands. Pointing at something will tell you how far away it is. However, the designers may want to reconsider their repertoire of gestures. As currently implemented, you locate another person by shaking your fist where you think they might be.

Parents of teenage gamers are often both bemused and concerned when they see their children twitching around like demented robots, utterly lost in some violent alternative universe. The effect will be even more striking as we move to a world of eye-tracking glasses, Leap Motion controllers, and even heartbeat-controlled music-making.[155]

The acquisition of virtual reality goggle–maker Oculus VR by Facebook for $2 billion confirms that major companies are betting on the future of combining data from both real and virtual worlds.[156]

As every place from airports to shopping malls to historic buildings starts using augmented reality technologies, we may face a new digital divide that will put non-users at a disadvantage. People without the right high tech eyewear may not receive important information. They might head to the wrong airport gate, miss a sudden flash sale at a store, or fail to appreciate the subtle architectural details of a building.

All the sophisticated display systems in the world would be useless without a solid stream of data to feed them. Data is being gathered at unprecedented rates, from all over the planet, as well as from space and with manned and unmanned aircraft, better known as drones. These will start to proliferate over domestic airspace as the U.S. Federal Aviation Administration issues new rules for their use. In *Drone Warfare: Killing by Remote Control*, Medea Benjamin notes that the infrared camera on a Predator unmanned aerial vehicle (UAV) "can even identify the heat signature of a human body from 10,000 feet in the air."[157] You can pretty well assume that if someone has one of these aircraft in your vicinity and wants to find you, they can do it.

Unless you live in a war zone, or near a border that is patrolled by air, you might assume that nobody is flying infrared sensors over your house. However, as thermal imaging technology becomes cheaper and more widely available, all sorts of projects are becoming possible. For example, the Slough Borough Council in the U.K. flew a thermal imaging camera over their Berkshire town and discovered 6,350 suspicious buildings that might be "sheds with beds"—possible illegal rental conversions.

A Calgary, Alberta-based energy project called HEAT (Heat Energy Assessment Technologies) has been flying a thermal infrared camera over homes in the area and posting the results online at www.saveheat.co. Anyone can look up their address, or anyone else's home, to see how they stack up in terms of wasting heat.[158] This information is provided openly on a website without any form of registration being required.

Data like this is a gold mine for contractors looking to get energy-saving business, for police looking for marijuana "grow ops," and for nosey neighbors. The project's organizer assures the public that "it is not our role" to report suspicious activity and that, upon request, you can have your home removed from the database. He also says that no one has ever requested to have their home removed, which would cause it to display as a blank space with the notation "PRIVACY CONCERNS."[159] Doing that would undoubtedly raise a red flag that something nefarious is going on in your home—what do you have to hide?

Even if the government is not trying to nab you for growing pot, you might face the wrath of civic and environmental organizations, or even citizen vigilantes looking to find "energy wasters" and shame them publicly.

While there is no reason to doubt the noble intentions of this project there is really nothing to prevent overzealous law enforcement from using the data collected in this manner. In a 2001 case, the United States Supreme Court ruled that using Thermovision imaging to detect the heat emitted by indoor farming was unconstitutional. However, it was a close (5-4) decision that could have gone either way. In his dissent, Justice Paul Stevens wrote that "Heat waves, like aromas that are generated in a kitchen, or in a laboratory or opium den, enter the public domain if and when they leave a building."[160]

An even more ominous aspect of the Supreme Court's decision in this case is that the unreasonableness of the search hinges on the use of "a device that is not in general public use, to explore details of the home that would previously have been unknowable without physical intrusion." However, how long will it be before our smartphones have thermal imaging and other intrusive capabilities?

The answer is "quite soon" according to Wilsonville, OR-based FLIR Systems, Inc., which is already taking orders for a sub-$350 add-on that turns an iPhone 5 or 5s into a thermal imaging device. Should this product succeed, peeking at your home with thermal imaging may fall into the "general public use" category.

Most people associate drones with large, military-grade price tags. However, every summer for the past few years at the DEF CON conference, hobbyist drone makers have shown off their latest homemade UAVs with the ability to intercept your cell phone signals and Wi-Fi traffic. Made mostly of Styrofoam, and with password cracking and other penetration capabilities built in, the manufacturing cost of the 2011 version was about $6,200.[161]

The use of aerial surveillance by law enforcement has seen a "function creep" from being used only in exceptional circumstances to being quite routine. According to a German research paper, police forces often feel justified in using drones in public places like sporting events because the fans have already consented to some degree of surveillance by entering the stadium and passing a warning sign.

However, Peter Ullrich and Gina Rosa Wollinger argue that drones are really closer to covert surveillance than normal security cameras: "Drones are quiet, fly high, and can even be used at night time, if equipped with infra-red or thermal imaging cameras. All these decrease direct visibility and therefore the possibility for the affected persons to realize their being under surveillance, to act accordingly, or just to be able to calculate the consequences of their actions."[162] The average person seeing a small aircraft, manned or unmanned, flying over has no way of knowing what data it is collecting, and who is going to use it for what purpose.

One of the most revelatory tales of retail tracking and datamining is described by Charles Duhigg in a *New York Times* article called "How Companies Learn Your Secrets." He explains that Minneapolis, MN-based Target Corporation created a model to detect, as early as possible, when a customer knew she was pregnant. This would allow the timely marketing of high margin baby-related products. The model was created by a statistician named Andrew Pole.

"About a year after Pole created his pregnancy-prediction model," Duhigg reports, "a man walked into a Target outside Minneapolis and demanded to see the manager. He was clutching coupons that had been sent to his daughter, and he was angry, according to an employee who participated in the conversation.

"'My daughter got this in the mail!' he said. 'She's still in high school, and you're sending her coupons for baby clothes and cribs? Are you trying to encourage her to get pregnant?'"[163]

It turns out that Target's pregnancy-prediction algorithm had become aware of this young lady's condition, from her purchases, before her father. He called the store manager a few days later with an apology. "It turns out there's been some activities in my house I haven't been completely aware of. She's due in August."

Duhigg quotes a Target corporate executive as saying "we found out that as long as a pregnant woman thinks she hasn't been spied on, she'll use the coupons. She just assumes that everyone else on her block got the same mailer for diapers and cribs. As long as we don't spook her, it works." So, to avoid the creepy factor, Target started mixing in photos of things like lawn mowers and wine glasses with the baby items in its mailings to mothers-to-be.

It is easy to imagine how this retail-tracking technology might evolve in the near future. Point-of-sale terminals could have cameras and sensors that read a customer's body contours. Pregnancy would be easy to detect in this fashion, along with race, ethnicity, and body mass index. All could be used to target the customer with advertisements and coupons for the next visit.

Peering further into the future, simply touching the keypad of the point-of-sale device might leave behind enough skin cells for a DNA sample to be collected.[164] This could lead to genetic profiling, tied to the customer's "Guest ID," which is already on file. Now the store has a wealth of data, not just on the customer but also on close relatives. Based on what stores like Target are already doing, there is every reason to expect that they would adopt technologies like this, unless they are made explicitly illegal.

Often, we enable tracking of our spending habits in return for discounts or points on loyalty cards. For example, two thirds of Canadian households have an active AIR MILES loyalty account. The points accrued on this can be used for everything from appliances to travel, and you are going to buy groceries or hardware anyway. The parent company of LoyaltyOne, the firm that runs AIR MILES, is Texas-based Alliance Data.[165]

Once I called the AIR MILES folks and asked them how much they know about me if I use their card at Safeway.

"We know you spent $35.62 and got two AIR MILES."

"So you don't know what I bought?"

"Nope, but if you also use your Safeway loyalty card they know exactly what you purchased right down to the SKU." In speaking to high school groups, I usually explain this with "they know you bought corn, cantaloupe, and condoms" which seems to make the point.

Somc might suggest dumping those cards, but they really have become *de facto* mandatory. Without the Safeway card, and its relatives at drugstores like Shoppers Drug Mart, Boots, Walgreens, CVS, and Duane Reade, you will wind up overpaying for your purchases compared to the person next to you.

Often a kindly cashier will offer a "courtesy discount card or code" to overcome your lack of a card. Since the cards are free, they do not want to bother having you apply for one to get the discount. Safeway's CEO, however, felt differently, ruling that no loyalty card meant no discount.[166] Anecdotally, there are tales of cashiers swiping their own cards and benefiting from the points, even taking long vacations with them, though companies are now tracking that.

Grocery purchases are not the only data being analyzed by machines. Now online job applications are also being screened in this fashion, such as the application to work at a Xerox call center.

Some of the factors used by Xerox to predict if an employee will stay on the job are obvious, like living close to the workplace. But there are also stranger indicators. According to an article in *The Economist*, "people who fill out online job applications using browsers that did not come with the computer (such as Microsoft's Internet Explorer on a Windows PC) but had to be deliberately installed (like Firefox or Google's Chrome) perform better and change jobs less often." There are even some findings that *The Economist* calls counter-intuitive. One example: "firms routinely cull job candidates with a criminal record. Yet the data suggest that for certain jobs there is no correlation with work performance. Indeed, for customer-support calls, people with a criminal background actually perform a bit better."[167]

Consumers can turn technology against merchants. You will often see people in stores cross-checking prices with online retailers and snapping photos for comparison purposes. As object recognition and visual search engines improve even more, you will soon be able to snap a photo of a passing car, or somebody's dress, and find it, at the best price, online.

Still, the technology deck does seem to be stacked in the favor of corporations over consumers. A Canadian IBM employee, Nathalie Blanchard, had her disability payments revoked when her employer's insurance company saw her online photos and decided she was having too much fun for a person on paid leave for major depression.[168]

Manulife Financial stated that it would never cut off a claimant solely because of information posted on social media, but admitted it is a source that they do consider. When Blanchard posted photos of herself frolicking on the beach, she certainly wasn't expecting them to be seen by her insurer. There is some debate about how private or public her Facebook profile was supposed to be, but it is probably not a good idea to "friend" your insurance company.

Your friends may even have an impact on your personal credit worthiness. Start-up loan site Lenddo is open about the fact that they allow prospective borrowers to "use their social connections to build their creditworthiness and access local financial services." This translates into a "Lenddo score from zero to 1,000," which they say "measures character." It is based on an undisclosed combination of "social data, information from your community, and data related to your Lenddo products."

One thing is clear: you want to be careful who you hang out with online if you want a high Lenddo score. On its website Lenddo acknowledges that "You should be selective when adding members to your community. Members of your Trusted Connections should be people that you know and trust. Their LenddoScores are derived from their social data, payment behavior with Lenddo products; most importantly they impact your LenddoScore as their LenddoScores increase and decrease."

If all your Facebook friends are deadbeats, you will be looking at a very high interest rate, or no loan at all. What if you try to stiff Lenddo? Let's just say they reserve the right to let your "Trusted Connections" on social media know all about it. And as your Lenddo-Score publicly plummets, you are dragging the scores of your friends down too.

Does social media shaming work? An employee at a bar in Reno, Nevada, got so angry when a customer ran out on a hundred-dollar bar tab that he snapped a photo of the guy and it was posted on the bar's Facebook page as a warning to other businesses that he was a "dine and dash artist."

The scoundrel wound up being arrested for other charges, and he promises to pay the bar tab someday—once he is finished serving his time. The story has gone viral, bringing the bar attention from people who will probably never set foot in Reno.

What you post and who you hang out with on sites like Facebook may hurt your job prospects or get you rejected during school admissions. Six U.S. states have passed laws banning employers from demanding the social media passwords of employees or prospective employees. If you are hoping for admission to an Ivy League school or a new job, and you have posted naughty things on Facebook, Twitter, LinkedIn, Tumblr, or Vine, you could of course just try to get rid of your online presence. But that is nearly impossible, say the experts. For one thing, you will live on in the postings of everyone you know. As Alan Katzman pointed out at *Business Insider*, "some (college applicants) have opted for a full social media lockdown or have simply changed the name on their Facebook profile. The risk of this approach is that colleges could rightly conclude that the lack of a social media presence means the applicant has something to hide."[169]

How you appear on social media can even influence whether or not someone will buy you a free lunch. Webinars have emerged as the dominant way to "educate" people about the finer points of new technologies. Of course they are usually "sponsored" which means you get to learn a lot about some vendor's products. To get you to attend, they usually offer incentives, often a trinket like a flash drive or USB charger for your car.

IT industry firm Condusiv Technologies raised the bar in October 2013, offering to buy your office a pizza in return for listening to their online pitch. They even offered to "include gratuity for the delivery person so there is no cost to you."

Since Condusiv wouldn't want to send pizzas to people who do not appreciate the subtleties of "I/O Optimization Techniques," they included a disclaimer reserving the right to refuse a pizza to anyone. But how would they tell a CIO from a janitor, when both might have a company email address? "We use LinkedIn to audit job title reflecting IT responsibility," they explained in their invitation email. If your profile says your skills include "Dinosaurs" or "Towels" or "Medical Marijuana" (real examples from a collection assembled by Mashable), there will be no pizza for you.[170]

So, before you solicit gag endorsements for Embroidery or Guinea Pigs or Dangerous Drugs (other real examples from that collection), you might want to think about who can see it. Information that you provide in one context, even in jest, can be used to form judgments about you in totally different and unanticipated situations.

Sometimes you do not even have to provide the information. It is accumulated automatically and shared without your consent or knowledge. When I first heard about Zoominfo, an online aggregator of information about people, I immediately went there and found myself. It showed my current professorial position, and even reminded me of some old projects that I had completely forgotten about. However, it also named me as a director of a Virginia-based Aerospace company and said that I taught at Bard College. Neither of these claims were true, so I claimed the profile in 2008 and corrected it.

I decided it would be fun to see what information they had back then on the Prime Minister of Canada, Stephen J. Harper. I was astounded to see him listed with the title "Campaign Director."

Campaign Director is a long way from Prime Minister of Canada. Zoominfo was using old information, and Harper had not yet "claimed" his profile, which you do by providing Zoominfo with a credit card that matches your profile name. They do not charge the card, at least for their basic service, but they do use it for identity confirmation. I mentioned Harper's profile in some Canadian government circles and soon his profile was claimed and updated.

Another under Mr. Harper's "Employment History" reveals how Zoominfo "thinks." Some news reporter, or blogger, apparently wrote about "Stephen Harper, the somewhat reluctant leader of the Conservative" party and the site's "patented" technology dutifully used that as his official title. It gets even worse. I logged on once and found him listed as the "Odious Leader of the Conservative Party."

You can actually use Zoominfo to play an interesting game of "find my doppelgänger." Aside from the well-known Tom Keenan at Bard College, I now know that other Tom Keenans run a beverage company in Portland, a trucking company in Illinois, and a foundation in Australia.

There is a Tom Keenan who teaches at an elite private school in New York City. Not only do we share a name, our email addresses are very similar, and so I know a disturbing amount about his private and professional life. I frequently get emails of the "dog ate my homework" variety from his students. I did finally have to call him when the Parents' Association at his school erroneously published my email as the contact for an upcoming "career day." I had Park Avenue neurosurgeons and Wall Street tycoons mailing me their confidential resumes as potential guest speakers.

There are three basic steps to handling information: input, processing, and output. So far, we've looked at how some creepy information systems suck up information about us as their input and process it, often to our detriment. Yet the creepiest technologies may be the ones that output their data, directly into our bodies and our minds.

# Sensation Creep

Walk into Kelly's Steak House in Las Vegas and you might find yourself instantly hungry for one of their signature dishes. Their secret? A frying pan full of delicious smelling, though probably inedible, onions, mushrooms, and spices simmering near the maitre d's podium. Realtors running open houses routinely toss bread in the oven to put you into that "let's buy a cozy new home" frame of mind. Shopping malls that are plagued with loitering teenagers have been known to chase them out with piped-in classical music.

We usually think of these tactics as just clever business practices. However, science is being pressed into service in whole new ways, usually to get us to buy or at least crave something.

McCain, the dominant producer of all things potato in Canada, has infested a series of U.K. bus shelters with gigantic baked potatoes. According to a report in *Advertising Age*, "a hidden heating element warms the fiberglass 3-D potato and releases the aroma of oven-baked jacket potato throughout the bus shelter. The aroma was developed over three months in collaboration with a specialist scent lab."[171]

Specialist scent lab? Actually, there is nothing new about that. In the 1970s, I was sent to interview an elegant executive at the posh Manhattan offices of Charles of the Ritz. They had just introduced a home accessory called "the Aromance Aroma Disc Player." It accepted little round "scent discs" and, using heat and a fan, filled your home with aromas like "Romance," "Fireplace," and "Movie Time," which smelled like buttered popcorn. Our chat was going along predictably until something provoked me to ask if they were working on any offbeat smells like, say, "Wet Dog." A pert wiggle of Ms. Charles-of-the-Ritz's shoulders told me this was the logical end of the interview. This product only survives as a nostalgic posting on the Internet.[172]

As we now know, they were definitely on to something—a creepy technology to get inside our minds. Neuroscientists say that the sense of smell is extremely effective at evoking memories and emotion. Anyone who has packed up the effects of a deceased parent has probably experienced the power of smell. The effect even has a name, the "Proustian Phenomenon," after the passage near the start of Marcel Proust's *Remembrance of Things Past* where the protagonist dips a cookie into tea, and a rush of childhood memories spews forth for the next 3,000 pages.

We are often unaware how we are being manipulated by scents until we catch ourselves moving over to the more expensive products or walking like a zombie into a restaurant we had planned to pass by.

Scent manipulation even shows up in funeral homes, which use a special industrial-strength cinnamon spray for odor control around decomposing bodies. A blog post by Sabine Bevers also reveals other tricks. She claims that Apple doctors its product packaging to emit a standardized "new device" smell, regardless of the product that is packed inside. Holiday Inn reportedly alters the sensory environment for different kinds of function bookings—pumping rose scent into the air for weddings and leather for business meetings. Bevers exposes a Brooklyn, NY grocery story for piping bogus bread smells into the air, and suggests that casinos use different scents to attract their preferred clientele. She also claims that "Nike stores use a mixed flower scent to direct you towards the more expensive shoe designs."[173]

Aromasys is one of the top companies in the "scent marketing" business, with offices in London, Hong Kong, Australia, and Las Vegas. Their website is discreet about their customer list, though they do trumpet one U.K. nightclub that was infused with "the scent of Watermint ... filling the club with the tantalizing smell of Mojitos" and that "the scent of chocolate was also in the air at the candy-filled reception held afterwards at a Hollywood Studio."[174] The movie was *Charlie and the Chocolate Factory.*

Australian researchers have discovered that at least one odor, freshly mown grass, can act directly on the hippocampus and the amygdyla to lower stress levels and improve memory. They have even founded a company, Neuro Aroma Laboratories, to market "Serenascent," a kind of "eau de grass" product based on their research.

"Serenascent was first conceived during a relaxing visit to Yosemite National Park in the United States of America," they write on their website. "The fresh calming aroma coming from the pine trees, sequoia trees and meadows inspired the inventors of Serenascent to study the relaxing effect of plants."[175]

While this mixture of plant-derived hexanals, hexenols, and pinenes has not replaced Valium yet, there is certainly anecdotal evidence that people feel a lot calmer after gardening or even mowing the lawn. If, on the other hand, your goal is to stir things up, especially in the erotic dimension, there are a number of human pheromone sprays that claim they will have that effect.

The Disney Corporation is famous for its use of sensory manipulation to create just the right atmosphere at their theme parks. One is the "Smellitzer"—a device that shoots out carefully engineered scents the way a howitzer sprays bullets. Technically it is "a method for sequentially directing at least two different scents from a gaseous scent-emitting system," according to a U.S. patent issued in 1986.[176]

You can smell the device in action at Disney parks, where the Smellitzer ensures that the Haunted Mansion is suitably dank and musty, and that the Pirates of the Caribbean ride evokes the smell of the sea. According to an article by Gabriel Oliver, the Smellitzer has a range of two hundred feet and there are fans to suck out the current smell when it is time for the next one to appear.[177]

Oliver notes that scent marketing also plays a role in other areas of the park. It helps to explain the sudden unexplained craving you get as you walk down Main Street, U.S.A., because "you're going to smell freshly baked cookies, whether there are cookies in the oven or not, thanks to the Smellitzer."

The Barclays Center in Brooklyn has reportedly come up with a signature aroma that is piped in to stimulate the appetites of those heading to Brooklyn Nets games and other events there. According to a local blogger, their smell technology comes from a company called ScentAir, whose motto is "Add More Excitement to Your Crowd Experiences."[178]

While many of the companies that use scent marketing are coy about it, the operators of the arena used by the St. Louis Rams acknowledge that they use scent marketing in their facility. They set out to answer a question that nobody had asked before—"what would happen if you pumped a cotton candy scent into your stadium?" The answer, says the ScentAir website, is a boom in the sale of cotton candy. However, they also saw sales of other food and drink items increase, because, says ScentAir, "the cotton candy scent triggers a response to buy food and drinks in general."

Scent marketers know that environmental aromatherapy can sneak into our consciousness, unearth treasured memories, and unleash retail spending, all at the subconscious level. Most consumers are blissfully unaware of this whole aspect of store design, and you almost never see warning signs posted about scent marketing. Who knows where else merchants, museums, and perhaps even governments are using scent to get directly to your deepest, most intimate feelings? Take a good whiff the next time you are standing in line at the post office or motor vehicles office. Is that a tree outside? Or perhaps ... Serenascent?

City University London professor Adrian Cheok feels that the Internet, as we currently know it, suffers from sensory deprivation. He is working on devices to transmit a person's body odor and body temperature, as well as sending tastes and smells electronically.

He plans to use coils in the back of the mouth to magnetically stimulate the "region of your brain that makes you perceive smells and tastes."[179] According to the report in *Motherboard,* Cheok is "working with the fourth best restaurant in the world (Mugaritz, in

San Sebastián, Spain) to make a device people can use to smell the menu through their phones."

Our other senses are also subject to manipulation, including the kinesthetic one that tells us which direction to walk for the lingonberry jam or the futons that are on sale. IKEA stores exhibit an uncanny awareness of human retail psychology. Long aisles take you past every possible piece of furniture, interspersed with handy gadgets you did not know you wanted. It is all specifically designed to keep you on the premises and shopping for as long as possible.

In a strange trance, many shoppers appear to completely surrender their free will when they enter the Swedish retailer's realm. Alan Penn, director of the Virtual Reality Centre for the Built Environment at University College London, has been quoted as saying that "the trick is that because the lay-out is so confusing you know you won't be able to go back and get it later, so you pop it in your trolley as you go past."[180] It is true that IKEA stores now have shortcuts. Those are, according to Penn, much more about fire code regulations than enhancing shopper convenience.

Our ears are not neglected when it comes to sense advertising. Blogger Bevers notes that McDonald's works "a sound that you could describe as something in between a deep fat fryer and a car heating system" into radio ads that run around lunchtime. The restaurant chain is also a great example of the use of instantly recognizable "sound logos" such as their five-note "I'm Lovin' It" jingle. Audio signatures can stretch much longer than five notes. United Airlines leaves its passengers humming "Rhapsody in Blue" and even pumps an "otherworldly" version of it into the tunnel connecting terminals at O'Hare airport.[181] Your smartphones and computer almost certainly pay a distinctive sonic tribute to their maker every time they start up.

Companies are even trying to brand certain tastes as being exclusively theirs. A guide to using taste in marketing campaigns notes that "Umpqua Bank used to buy a few dozen cases of honey bottles from a local honey farm each summer, and put its own label on the bottles as part of their bee-themed Summer Swarm promotion."[182]

In *Salt Sugar Fat: How the Food Giants Hooked Us*, Michael Moss describes how what he calls "the creators of crave" engineer their products to exploit human psychobiology. Writing on a *New York Times* blog, Moss explains how Taco Bell's Doritos Locos Taco is "a marriage made in processed-food heaven."

"It has dynamic contrast," he says, as well as great mouthfeel and a lingering smell that stimulates food memories and cravings. "But yet," he says, "this is the strangest one of all, it's forgettable. None of the many flavors of Doritos Locos Tacos are strong enough to trip a signal called 'Sensory Specific Satiety,' that will cause you to feel like you've had enough. That's the point. They are designed to make you want more."[183]

Indeed, Arch West, "the Frito-Lay marketing executive credited with inventing Doritos," was buried with the wildly popular snack food that he created.[184] Mr. West made it to the age of 97 and died of natural causes. Will you be so lucky?

# Bio Creep

Would you want to know the exact date and time of your death? It is a thought-provoking question with disturbing implications. A number of death calculators are available online, each providing a disturbingly precise date for your demise. Some even show you your future tombstone. One of my favorites, www.deathtimer.com, uses age, gender, geographic location, body mass index, and smoking/drinking habits to form an educated guess.

I had to speculate a bit about the details of a certain well-known former intelligence contractor, but the result did have a creepy ring to it, particularly because this site claims to base its predictions on "life expectancy statistics from the CIA, United Nations, and other sources." Edward Snowden's predicted date of demise is July 12, 2060, and his epitaph reads "Russian roulette isn't as fun as it appears."

While Internet death calculators all come with "for novelty use only" disclaimers, there are an increasing number of tests that really can provide actual scientific information about your probable life-span. Tests indicating certain BRCA1 and BRCA2 genetic mutations have led actress Angelina Jolie and many other women to undergo preventative double mastectomies to avoid potential breast cancer. Cholesterol and PSA tests, and what you do about the results, can potentially affect your longevity. Genetic screening tests can also help you plan a family, and possibly even help you engineer your children through pre-implantation genetic diagnosis. These tests can raise life-altering and disturbing ethical questions. You may not want to open your test results when you get them, and you certainly do not want to share them too widely.

The advent of HIV testing in the mid-1980s brought the privacy aspects of medical testing squarely into the public's consciousness. As the seriousness of the disease and its routes of transmission were discovered, there were calls for the mandatory testing and disclosure of results to protect others. There have been a number of high-profile court cases where someone who is HIV-positive has knowingly infected someone without providing proper disclosure.

Two Canadian men, Stephen Boone and Noel Bowland, were each sentenced to jail terms in 2013 for not disclosing their HIV status to partners. Boone was also charged with administering a noxious substance—his own semen.[185]

Public health authorities want people to be tested for HIV, and the medical nature of that disease has changed from an almost certain death sentence to one that is more manageable. Most civilized societies follow the World Health Organization guidelines for HIV testing. These procedures require that the testing be done only with informed consent, that counseling be made available, and that the test results be kept confidential. In some parts of the world, however, this standard is ignored.

The Guangxi Zhang autonomous region in the south of China now requires real name registration for HIV testing, in the interest of controlling the region's HIV rate, which is one of the highest in the country. The legislation, which took effect July 1, 2013, also requires those who test positive to warn their spouses or partners.[186] Test results would still remain confidential, though how much people can rely on such a guarantee remains to be seen.

In Canada, the policy director of the British Columbia Civil Liberties Union has expressed concern about that province's "Seek and Treat" campaign, which seeks to prevent and treat HIV infections. Calling it "a perfect storm of privacy concerns in relation to HIV testing," Michael Vonn says that there is now a push to make HIV testing routine, and to erode the privacy associated with testing. He notes that in Vancouver, HIV test results were once only delivered in person. Now, negative results can be obtained over the phone, which Vonn notes "is tantamount to giving all test results over the phone." He also cites the province's "newly instituted data-sharing systems providing broad access to personal health information along with legal reforms that allow for that broad access."[187]

With undoubtedly good intentions, the Huntington Society of Canada ran televised public service announcements in 2013 suggesting that the simple act of being tested for Huntington's disease might open you up to discrimination in Canada.[188] The commercial's dramatic statement "being tested puts you at-risk" was actually somewhat misleading. Having the test itself is not the problem. It is how the results are handled that requires some careful consideration, both on an individual and societal basis.

The Huntington Society explains that Canada is the only G8 country that does not have legislation banning genetic discrimination. The group is spearheading a coalition to push for legislation forbidding discrimination based on genetics in Canada.

Current Canadian law provides a few after-the-fact remedies for people who are discriminated against for genetic reasons, but certainly not the same level of protection as in the U.S., the U.K., and most countries in the European Union.

The U.S. Genetic Information Nondiscrimination Act of 2008 prohibits genetic discrimination in employment and health insurance, although life and disability insurers are still generally allowed to base their decisions on whatever genetic information they can get their hands on.

In a 2012 report, the Privacy Commissioner of Canada surveyed the global situation on the use of genetic information by insurers. They found that, in the U.K., "the Association of British Insurers and the Government have agreed on a voluntary moratorium, recently extended to 2017, on the use of predictive genetic test results for life insurance policies under £500,000 or critical illness policies under £300,000." In Germany, "insurers may only request genetic test results for life insurance policies that pay out more than €300,000 or for disability policies that pay more than €30,000 annually."[189] These results were noted in a study by the Privacy Commissioner of Canada which concluded that "the issue is far from settled in Europe."

Huntington's disease is a particularly fascinating kind of genetic lottery because if one of your parents has the disease, you have a 50/50 chance of inheriting it.[190] There is currently no cure and people who have it ultimately die from complications like pneumonia and heart disease.

Before genetic testing, children at risk for the disease lived under the shadow of getting it later in life. Now that it is possible to test for this abnormality, even *in utero*, new ethical problems present themselves. Is abortion of a fetus that is likely to be affected ethical? Given that there is no cure, or even any kind of really effective treatment, at what age should a person be tested? What good will it do an eight-year-old to know that the future holds a currently incurable disease?

The Huntington's test can make predictions with virtual certainty because the disease has autosomal dominant inheritance which is well understood. If you have more than 41 copies of a certain section of DNA on chromosome 4, you will almost certainly get the disease.[191] Having 35 or fewer copies is not usually associated with the disease, and those with 36 to 40 copies have what is called "incomplete penetrance" with increased risk of having or passing on the disease.

While the Huntington's test can usually give a definitive answer, many other genetic tests are predictive, providing an indication that a person is at higher risk for a specific condition. We are told with some regularity that scientists have found a gene for alcoholism. Usually these results are less than convincing, especially since they typically deal with alcoholic mice or rats.[192] Also, it's clear that someone who never touches a drink will not become an alcoholic, though they might express their addictive tendencies in other ways.

Since everyone has to die of something, many people find the idea of being told what it might be unnerving and unnecessary. Of course, the counterargument is that you might be able to make more informed lifestyle or treatment choices with better information. Knowing your genetic heritage might also be a factor in deciding whether or not to have children, and even a consideration in your choice of partner.

The Academy Award–winning filmmaker John Zaritsky has made a documentary called *Do You Really Want to Know*? about the ethical side of genetic testing. He notes that there is a history of Alzheimer's disease in his family, but that "in my case I've resisted being tested because I don't think my life will be improved any by the knowledge."[193]

As Zaritsky explores in his film, once you have a test and choose to view your results, you can never turn back the clock. He does find that some people have made good use of their positive Huntington's test results. It has helped them plan their lives, and in one case, a couple where one spouse had the gene was able to make use of pre-implantation genetic diagnosis to select healthy embryos, sparing them the heartbreak of having children with the disease.

However, as the universe of possible genetic tests expands, we might soon be bombarded with television commercials suggesting we are somehow derelict in our duty, to ourselves and our families, if we do not probe our genetics and take action on it.

Speaking at the 2013 meeting of the American Association for the Advancement of Science, Brian McNaughton, founding scientist of the direct-to-consumer genetic testing company 23andMe, frankly assessed the state of our knowledge of the human genome, saying that "Almost everything in the human genome is a variant of unknown significance." This does not stop his company from offering Internet-accessible $99 genetic reports to anyone willing to pay. However, the U.S. Food and Drug Administration (FDA) did put a crimp in their business model. On November 22, 2013, they ordered the company to cease providing health-related information to customers. The firm will still run your test, and provide raw genetic data, which you can then take to other sites (such as OpenSNP.org) for interpretation. It is certainly not as convenient as the genetic risk reports that 23andMe used to provide.

Why is the FDA so reluctant to let you see your own genetic data? Information like this used to come through a medical professional, presented with suitable interpretation and caveats. The FDA is clearly concerned that some people may base important medical decisions on such test results, which are presented without the intermediation of a medical professional.[194] The company is looking forward to the day when they can resume this line of business. On their website, 23andMe notes that "In the future, you may be able to receive health-related results, dependent upon FDA marketing authorization."

Lots of smart, tech-savvy young people are using direct-to-consumer testing to investigate their genetic makeup, often before starting a family. They treat the results as useful information, but not as a definite roadmap of their medical future.

The smartest ones do the test under an assumed name and pay for the results in some anonymous fashion. However, even that precaution may not be enough. Technology expert Brian Lynch explains that "people can tag me as a sibling or cousin, and then share THEIR results to others, so they're in effect sharing some of 'me' to the public."[195]

Even with the U.S. Genetic Information Privacy Act in force, letting anybody have your DNA linked to your real identity is like handing them the keys to your body, as well as information about all your close relatives.

In 2005, IBM announced that they would take a leadership role in promoting the cause of genetic privacy. The tech giant pledged not to use the genetic information of employees or prospective employees in making decisions about them. Then-IBM president Sam Palmisano explained the logic behind this policy: "It has been IBM's long-standing policy not to discriminate against people because of their heritage or who they are. A person's genetic makeup may be the most fundamental expression of both."[196]

While 23andMe gives some assurances in its privacy policy, they specifically reserve the right to disclose, to third parties of their choosing, your "Genetic and Self-Reported Information that has been stripped of Registration Information and combined with data from a number of other users sufficient to minimize the possibility of exposing individual-level information while still providing scientific evidence."

Even in anonymized form, this data will be gold to pharmaceutical companies who may use it for research, but may also be putting it to less noble uses. Genetic testing companies are subject to court orders to disclose information, hacker attacks, physical break-ins, and the misuse of information by their own employees.

Unlike a credit card that you can cut up to get a new number, or even fingerprints that you might burn off, you are not going to change your DNA in this lifetime. Sometime in the future, someone may create a de-anonymizing system that could pick you out of a genetic crowd, even with technical safeguards in place.

If there are only a small number of people with certain traits, and governments keep building giant DNA databases, then they have the tools and incentive to round up people who fit a certain profile to question them in the event of a crime committed by a person with this same profile.

It may or may not make you feel better to know that Google, a company whose modest mission is to "organize the world's information and make it universally accessible and useful," has invested millions in 23andMe. In addition, Google's co-founder Sergey Brin and 23andMe's co-founder Anne Wojcicki have had two children together.

23andMe is not without competition. In New York City, Jared Rosenthal is the owner of a mobile DNA testing truck, actually a repurposed recreational vehicle. He calls it a "24/7 drug testing, breathalyzer, and DNA testing operation."

Rosenthal told *Bloomberg Businessweek* that he started this business in an RV because he "couldn't afford to lease an office" but he now sees the virtue of a lab that invites the public to come in for answers to burning questions like "Who's Your Daddy?" for a fee of $299, or $349 if you might need to use the results in court.

On his company's website, Rosenthal writes that "We conduct hundreds of these tests each year—and we've had plenty of 'you are not related' moments." He also reports using genetic testing to connect a woman with her long-lost daughter, who she had been told died in infancy. He even tested a "married couple who discovered they shared—of all things—the same father." Tales like this cry out for a reality TV show, and, Rosenthal reports on his site, one is in the works.

I was told a fascinating story by a transplant surgeon at a major North American hospital. A man who needed a kidney transplant brought his twentysomething daughter with him to the hospital. When tested for compatibility, it was clear that she was not genetically his daughter. The hospital convened its ethics board which decided the doctors should tell the man, and give him the chance to break the news to his daughter.

He declined, so they informed her and she stormed out. A week later she was back, having re-considered the situation. "I'm going ahead with the operation," she said, "but I'm so glad that I know the truth. If I had found out later I would have felt somehow cheated."

The chances are good that when the young lady was born, her heel was pricked with a needle for a blood sample. Since the 1960s, this procedure has been routine, and even legally mandated, to test for a number of inborn errors of metabolism. The justification for Newborn Blood Screening (NBS) is that with the right kinds of treatments, serious diseases like phenylketonuria can be detected right after birth and managed, averting tragic medical problems.

Typically some of this blood is also put on filter paper, dried, and filed away. These samples are an excellent source of DNA. In fact, newborn screening may actually be the Holy Grail that many governments have been lusting after—a national DNA database of all of its citizens.

Now that people are realizing this hidden source of DNA already exists, Americans are doing what Americans do when they're upset—filing lawsuits. According to an article in *Pediatrics*, "Within the last five years, two states, Texas and Minnesota, have been sued by parents because of the retention of residual bloodspots without parental permission. Texas, in a negotiated settlement with plaintiffs, agreed to destroy ~five million retained specimens in 2010."[197]

We all want the benefits that come from genetics research, but we are skittish about having our own, or our children's, genetic data filed away somewhere. Knowing how it can be misused today, and extrapolating to future technologies, justifies our intuitive anxiety.

A survey of Canadians showed "strong support for storage of NBS samples for quality control, confirmatory diagnosis, and future anonymous research."[198] However, many of those surveyed also expressed discomfort. Some expressed lack of faith in the health system's security procedures. One noted that there was a powerful incentive for misuse and that "One of the corporations is going to pay for all these blood samples ... And they might be using it for different purposes." Another reported objection was that, in the future, "someone can request and through a court get a court order to go and get this genetic material that belongs to your baby. That I find I'm a little more nervous about."

Lending credence to these fears is the trend towards opening up adoption records. Many birth parents entered into blind adoption believing that the records would be sealed forever. Now organizations like the Adoptee Rights Coalition are fighting to give adopted persons the right to obtain their original birth certificates. DNA can provide the perfect key to unlocking things that people thought would be kept secret forever. Just ask sperm donors who are now starting to hear from their "anonymous" offspring.

The technology for producing the semen sample really has not changed much over the years. Small room, locked door, sample container, pawed-over magazines. Nor has the compensation, which one California sperm bank lists as "$100 per donation, up to $1200 per month." What has changed greatly is the screening process for prospective donors. It now includes family history, medical records, blood type, childhood photos, pets, your philosophy for achieving World

Peace—basically everything a Miss America contestant would need to disclose, all organized and indexed in an online database.

The website of Fairfax Cryobank in Fairfax, Virginia shows prices from $250 to $735 per dose, depending on, among other things, the educational attainment of the donor. Like any good retailer, they have an elite "Club Fairfax" that allows prospective mommies first crack at new sperm donors plus a "buy five vials and receive the 6th free" offer. There are no (publicized) auctions of donor sperm, though there certainly is a resale market. Sperm banks trade and sell semen to each other. This is a good thing, because you would not want all the kids in the same geographical area to be genetically related. The risk of accidental incest a few decades later is just too great.

This sperm bank makes selecting the father of your child a lot like ordering a pizza. You choose Donor Ancestry (Any, Asian, Black, Caucasian, Latino, Multi), Eye Color, and Hair Color. You can even upload a picture of someone you would like the father to resemble to help find the Dream Daddy of your not-yet-conceived offspring.

To test how well their matching algorithm performs, I uploaded a publicity photo of Brad Pitt (since the Internet assures us he is the "Sexiest Man Alive") and the Fairfax Cryobank system helpfully churned out two HIGH matched donors and five MEDIUM matches. Specifying that the sperm donor should also be Black produced only LOW matches, increasing my faith that this rather opaque matching system is doing some actual processing. We ultimately have to take the system's word for the quality of the matches since the majority of sperm donors choose to be anonymous. At least they think they are.

Consider the case of Donor 401 at Fairfax Cryobank. A 240-pound, 6'4" man of German ethnicity, his hobbies included football, hockey, and, apparently, sperm donation. The website donorsiblingregistry.com currently lists seventeen of his progeny, but there are believed to be many more out there. He even has eleven women in his fan club—all mothers of children he sired by artificial insemination. While he retired from active sperm donation in 2004, some vials of his sperm were laid away in storage. Ah, but do VIP sperm samples appreciate in value like a 1787 Château Lafite Rothschild? Actually, they can become priceless in a creepy sort of way.

There is apparently a waiting list for any available vials of seed from Donor 401 at the Fairfax Cryobank. In 2006, a California artist named Melissa Weiss realized she was sitting on a supply of seventeen vials that had cost her $175 each. Since she was no longer trying to conceive with donor sperm, she decided to do a very good deed. Journalist Phillip Sherwell, writing in *The Telegràph,* explains: "She insisted on giving them to existing '401 mums' who wanted more children by the same donor father. There seems little doubt that, had she chosen, she could have sold the samples for many times the purchase price."[199]

Nobody is really sure what the record is for number of children sired by the same sperm donor. But a report in the *New York Times* discusses a woman who started a kind of online playmate group for her son who was conceived through donor sperm. The membership roster is now up to 150 half-siblings and some of them go on vacations together.

"It's wild when we see them together," she's quoted as saying. "They all look alike."[200]

According to Barnard College President Debora Spar, author of *The Baby Business: How Money, Science and Politics Drive the Commerce of Conception*: "We have more rules that go into place when you buy a used car than when you buy sperm."[201]

Spar would probably be even more horrified by the totally unregulated sperm donor matchmaking that happens on social media sites including Facebook. Just search for "Sperm Donor" and you will find frantic posts like this one "Are there any donors in the New York, New Jersey, or Pennsylvania area. Will be ovulating in ten days."

So far, the famous Donor 401 has managed to maintain his anonymity. Will he ever be on the hook for college and wedding expenses for his global brood? Never say never. Beside a photo with the caption "Ready for a phone call in fifteen years?" *New Scientist* ran a fascinating story of a fifteen-year-old boy who tracked down his sperm donor father with a cheek swab and some detective work. He did this despite the fact that his father had never contributed DNA to FamilyTreeDNA.com, the genetic testing site that the boy used to test his own DNA.

As Alison Motluk writes, "the teenager tracked down his father from his Y chromosome. The Y is passed from father to son virtually unchanged, like a surname. So the pattern of gene variants it carries can help identify which paternal line an individual has descended from and can also be linked to a man's surname."[202]

Eventually he found two men with DNA that closely matched his own. They had the same surname, and he was able to use this clue, plus other known information such as the sperm donor's place of birth and college degree, to find his biological father.

As we see from the change in sperm donation secrecy over just a few decades, yesterday's ironclad secrets have become today's open information.

What will happen to today's secrets as technology moves forward?

# Body Creep

In 1998, a British professor named Kevin Warwick made history by becoming the first human to have an RFID chip surgically implanted into his body.

Warwick was the first "chipped human" for a period of time, until businesses saw the potential of this kind of body alteration. In 2004, a beach club in Barcelona started to implant chips the size of a grain of rice into their VIP customers. These RFID chips provided access to restricted areas and also served as identification when buying drinks, allowing customers to wear skimpy attire without needing pockets for ID or credit cards. The chip emits a unique ten digit number and, according to BBC producer Simon Morton, the implantation is painless.[203]

Kevin Warwick, the RFID pioneer, has also embarked on other groundbreaking biohacking adventures, for instance creating a mechanical hand that responds to impulses from nerves in his own arm. He also claims to have conducted "the first purely electronic communication experiment between the nervous systems of two humans."

In a speech at a computer conference, Warwick told a somewhat racy story about the time he was on one side of the Atlantic and his wife was on the other, along with the robotic hand. Details of what ensued are probably best left to the imagination. However, he was certainly a pioneer in a whole new field of computer applications with a tantalizing name—teledildonics. Warwick's gutsy work went a long way to preparing us for the idea that technology can, and perhaps should, be used to take us beyond the human body that we arrived in.

Personalized medicine, diagnosis and treatment tailored to a specific individual, is widely regarded as the "next big thing" in health care. In 2009, the U.S. Secretary of Health and Human Services, Michael O. Leavitt, gave a techno-optimistic appraisal of modern medicine in the foreword for *Genomic and Personal Medicine*, a textbook on the human genome: "As diseases come to be understood at a new level, we should be able to better achieve the right diagnosis and the right treatment for each person without the trial-and-error process that has long characterized medical treatment."[204]

Popular books followed, such as one by Kevin Davies that starts with the story of Dr. Jeffrey Gulcher, founder of deCODEme, the pioneer in direct-to-consumer DNA testing that started selling its reports for $1,000 in 2007.[205]

Running Gulcher's own cheek swab produced some predictable results such as a tendency to baldness, which was already evident on his head. But, Davies writes, "a handful of those DNA markers suggested that Gulcher had double the average lifetime risk for type two diabetes and prostate cancer."

Sure enough, a biopsy revealed prostate cancer at grade six on the Gleason scale. Gulcher credits his company's test with saving his own life.

By the time the 2013 edition of *Genomic and Personal Medicine* came out, Leavitt's essay was replaced by the writing of Dr. Eric D. Green, Director of the NIH's National Human Genome Research Institute.

Green showed much more restraint, noting that in health care delivery "the actual implementation of genomic and personalized medicine … is associated with myriad nuances and complexities that will take many years to appreciate fully and to address adequately."

Despite the subtleties of personalized medicine, people are already making important health care decisions based on genetic test results. Dr. Jay Orringer, Angelina Jolie's Beverly Hills plastic surgeon, trumpeted that her decision to opt for a preventive mastectomy "has already begun to save lives," by inspiring women to get tested for a genetic predisposition to breast cancer.[206]

Personalized genetic medicine is not as simple as "you have this gene, so do this." For example, it has been discovered that some people are much more sensitive than others to blood thinning drugs, and also process them differently. There are apparently genetic markers (polymorphisms in CYP2C9 and VKORC1) associated with these differences. The correct dosage of blood thinning drugs is important since doctors must tread the line between using too little, which can allow blood clots, or too much, which can cause internal bleeding. It is also complicated by differences among the races.

Research at the University of Pennsylvania School of Medicine and seventeen other hospitals tried using this genetic information to improve dosage calculation in over a thousand patients during their first four weeks of therapy. They concluded that having the genetic information was no better overall than the old way, and the outcome was actually worse for one subgroup (African American patients).

Our ability to measure and change our bodies using technological advances has outstripped our ability to reflect on the repercussions of what is actually happening. We have seen this before with reproductive technologies, such as the ability to screen embryos for *in vitro* fertilization: it was not long before some parents were asking doctors to select for a preferred gender at the embryo stage. Many found this to be a creepy form of tampering with nature. The ethics committee of the American College of Obstetricians and Gynecologists wrestled with this issue and decided that sex selection was acceptable to avoid having a child with a sex-linked genetic disease. However, they expressed opposition to "meeting other requests for sex selection, such as the belief that offspring of a certain sex are inherently more valuable."[207]

Our incomplete understanding of science and technology, coupled with our human desire to "do something," can lead us down some very dangerous pathways.

At the 2011 DEF CON hacker conference, I covered a press conference called by Jerome Radcliffe, a security researcher who is also diabetic. He uses an insulin pump. At the conference, he demonstrated how easy it would be to hack the device, with possibly fatal consequences. He did leave out a few key details in the interest of preserving his own life.[208]

Members of the non-technical press focused on the alarming and sinister aspects of this story, such as "a stranger wandering a hospital or sitting behind a target on an airplane would be close enough (to take control of an insulin pump)."[209] Yet there is a deeper, and in some ways creepier, aspect to what Radcliffe discovered and disclosed to the world. In the wrong hands, devices that are supposed to save us can injure us. If manufacturers continue to eschew encryption, and the devices leak data, highly personal information about us can travel to places it should not go.

Radcliffe's hypothetical attack on his insulin pump could have been thwarted, or at least made more challenging, by the use of encryption to protect the data going into and out of the device. However, encryption uses processing power that shortens battery life. The good intention of giving the user the longest period between replacements works against the goal of enhanced security. As technology moves deeper into our bodies, we will need to make decisions about trade-offs like this.

Barnaby Jack, a noted security researcher whose ideas about hacking cardiac pacemakers inspired a plot twist in the TV series *Homeland*, planned to present ground-breaking research about medical device vulnerabilities at the Black Hat 2013 conference. He was scheduled to tell thousands of people how to use "a common bedside transmitter to scan for, and interrogate individual medical implants."[210] Days before his scheduled presentation, Jack was found dead in his San Francisco apartment, in a death ruled as accidental due to "acute mixed drug intoxication."[211] Conspiracy theorists continue to suggest that he was murdered for what he knew, and was about to reveal.[212]

Formerly the domain of advanced researchers with well-funded labs, a product based on the NeuroSky biosensor now can be ordered by anyone on the Internet.[213] Billed as a research-grade EEG (electroencephalogram), the brainwave analyzer connects to your head with no drilling or other unpleasantness, and allows you to track your Alpha, Beta, Gamma, Delta, and Theta waves using a "non-invasive single dry sensor that reads brainwave impulses (not thoughts)."

The Mindwave comes with applications to help you focus, meditate, and do math in your head. Hobbyists have already had a field day with this device, rigging up brain-controlled robots, computers, and the ever-popular mind-controlled garage door opener. More adventurous types are using it to track lucid dreaming, where, as blogger Brian Benchoff put it, "you take control of your dreams and become a god."[214] One practitioner of this art claims he can communicate from inside his dreams by blinking in Morse code.

An even more powerful product is Myndplay, which allows users to engage with video and actually affect the outcome. The creators envision a movie theater full of people wearing their sensors; their hive mind will alter the plot in real time. In a promotional video for this product, a narrator with a British accent assures us that the audience will finally have power over creative works: "they choose the direction, they decide the outcome, whether they want to or not. Their minds will determine what happens."[215]

If you just want the world to know something about your state of mental arousal, you can purchase some Necomimi Brainwave Controlled cat ears. They stand up when you concentrate and lie down when you relax. And sometimes they just wiggle suggestively. This is what happens when a girl in the promotional video spies an attractive young man. He ignores her, and the ears flop down.[216]

Devices like these can certainly provide insights into what is going on inside our heads. However, we generally assume that what got into our minds is the result of real things we have experienced and real thoughts we have had. That assumption is starting to develop cracks.

There are some needlessly frightened mice running around at the RIKEN-MIT Center for Neural Circuit Genetics. Researchers there have made the rodents afraid of getting an electric shock. However, they had never actually experienced the shock. It was put there by interfering with cells in the hippocampus.

This study sheds light into how we build up memories in groups of neurons, called engram-bearing cells, and how easily they can be tampered with. As the researchers, led by Steve Ramirez, reported, "Our data demonstrate that it is possible to generate an internally represented and behaviorally expressed fear memory via artificial means."[217]

According to a press release accompanying the study, such research is vitally important because "almost three-quarters of the first 250 people to be exonerated by DNA evidence in the U.S. were victims of faulty eyewitness testimony."

It is also important because DARPA, the Defense Advanced Research Projects Agency folks who gave us the Internet, are keenly interested in false memory, under the euphemism of "narrative networks."[218] It fits well with what appears to be DARPA's somewhat creepy new slogan, expressed on its home page: "Creating and Preventing Strategic Surprise."

Has this federal agency suddenly decided to study *Harry Potter* or *Fifty Shades of Grey*? Actually, they are much more interested in how people become convinced to join terrorist cells. They awarded a research contract worth about $7 million to a group led by Charles River Analytics of Cambridge, MA. In a news release, Charles River noted that as part of this project, they will be developing a piece of software poetically named "SONNET, or Studies to Operationalize Neuro-Narratology for Effective Tools."[219]

They also noted that "Charles River will conduct neurological studies to understand what makes a story compelling, and develop tools to sense and forecast people's reactions," echoing a previous DARPA project where the brains of college students were monitored with functional Magnetic Resonance Imaging (fMRI) to try to predict what they would find funny in a TV sitcom.

Reading somebody's fMRI is still a somewhat cumbersome and invasive process, and not likely to be done at a distance any time soon, though conspiracy theorists claim the CIA does it all the time. A brain pattern called P300 is a lot easier to capture, and is definitely being used in military and intelligence settings. P300 is a measurable electronic event, called an "evoked potential" that occurs about 300 milliseconds after a stimulus, and is associated with recognizing this stimulus. The U.S. military believes that determining evoked potential will be very useful in helping soldiers rapidly recognize dangerous situations. They are using this approach in a set of smart binoculars called Sentinel (SystEm for Notification of Threats Inspired by Neurally Enabled Learning). Through what is being called a "brain-machine interface," these "cognito-neuro" binoculars tap into the unconscious ability of the soldier to identify threats, and appears to speed up the process by about 30%.[220]

The Jasons, a group of top scientists that advises the U.S. military on technology matters, has expressed some concern about this type of technology, noting "potential for abuses in carrying out such research, as well as serious concerns about where remediation leaves off and changing natural humanity begins."[221] And of course, if militarized brain technology falls into the wrong hands, as it certainly will, it can be used against us.

On a more peaceful front, investigators at Glasgow University have been able to use fMRI technology to decode the brainwaves of people who viewed images of "happy and fearful faces," looking for differences. They learned what every portrait artist knows, that the eyes and mouth are important indicators. Despite the early stages of this work, the end goal is to form a "scientific" opinion about what someone is thinking.[222] Some have suggested that in the future people may leave their memories, transferred onto suitable media, for their heirs or for researchers. Your estate might even get a tax receipt.

The tax deduction might even be determined by how interesting your life has been. There's a "profession of the future"—digital life appraiser.

We already have a good if creepy idea what your physical body is worth. A science blogger with too much time on his hands recently quantified the amounts of various chemicals in the average human body and linked them to market prices. He claims that a typical result for your component chemicals is approximately $1,985.77.

To allow for fluctuations in the value of your 140 grams of potassium and 780 grams of phosphorous, he has posted his Excel spreadsheet online so you can calculate your own value. This, of course, assumes you could extract and purify all the elements in your body and sell them at market rates.[223]

There is also a "Cadaver Calculator" that asks you twenty questions then tells you how much you are worth to medical science, for instance: "Congratulations, Your Dead Body is Worth $4165."[224]

But why bother doing those when the market already sets a price for a human body?

A report from Hyderabad, India, claims that "500 unclaimed or unidentified dead bodies were sold from the Osmania mortuary in the past two years."[225] This article also gives a rough price list with "Rs 20,000 (about $325 US) to Rs 30,000 for bodies of those who died in accidents and Rs 30,000 to Rs 40,000 for bodies of those who died of natural causes." It also notes that bodies of younger persons are more in demand.

The probable destination for these cadavers is private medical colleges in India. North Americans are much more discreet about this process, though their medical schools need cadavers too. Instead of relying on dedicated alumni to pass away and leave their bodies to their *alma mater*, many schools use the service of a "fee-based service organization" called Science Care.[226] From their website, it seems that many of the donated bodies wind up in pieces, used to test medical devices and for other purposes where human tissue is required.

Of course, organ donation is certainly the best and highest use of a human body, and something everyone should consider. To their credit, the folks at Science Care encourage organ donation and their site says "most of the time individuals can be BOTH organ donors for transplant purposes as well as whole body donors for medical research and education." Although they cost more than actual cadavers, and cannot be dissected, some medical schools are switching over to plastinated bodies for their anatomy labs.[227]

Anyone who has seen one of the human taxidermy exhibits, in which plasticized bodies are put on display, will know that there is another possible afterlife for your mortal remains.

I interviewed Dr. Gunther von Hagens, the originator of the process and the force behind the hugely successful "Body Worlds" exhibitions. He told me about the early days of developing his patented process, with bodies exploding in ovens. Unlike his competitors, von Hagens operates an active program soliciting body donations. His Germany-based website reports "about two visitors a day seeking inclusion in the Institute's body donation program."[228]

A medical student once told me a touching story about an elderly woman who wanted to give her body to science but was concerned it would not be treated with proper respect. She insisted that a note travel with her cadaver, saying "Dear Future Doctor: I hope you learn a lot from probing my body and will treat it with respect. I just thought I'd mention something. Remember when you thought you would fail that organic chemistry exam or not do well on the med school interview … well, I was praying for you then."

Sadly we cannot attach sweet notes like that to the data about our bodies as it gets passed around and monetized. It would not do much good anyway. How do you ensure that data about your body is treated with respect by an automated big data decision process, running on a corporate or government computer? Decisions about the fate of our digital *corpus* are made in nanoseconds with all the precision and coldness of algorithmic logic.

Our living bodies are always capable of providing one of the three factors ("something you have," "something you know," "something you are") for identifying ourselves to our technology. Biometric technology shows up regularly in Hollywood films. In one of Arnold Schwarzenegger's movies, *The 6th Day*, he spoofs a fingerprint-controlled biometric identification system with a cloned version of somebody's finger, but drops it in the process. He utters the predictable line "I'm all thumbs today." Having worked with Hollywood scriptwriters I am pretty sure that one got a laugh in the writer's room. Yet the collection of biometric data is very serious business, and something that many people intuitively feel they should resist.

Many years ago, in an attempt to counteract fraud, a New York-based bank started requiring a fingerprint from customers when they cashed or deposited checks. The bank quickly stopped using it when clients complained that they did not want to be treated like criminals. This Touch Signature® product is still being used by pawnshops and furniture rental outfits. According to a 2002 *New York Times* article, the Texas Bankers Association bought 80,000 of the inexpensive blue ink pads and sold them "to nearly half the banks in Texas and to banks in 37 other states." There are still countries where voters are fingerprinted and Princeton University researchers have noted that, while fingerprints on ballots may allow for better auditing of election results, there is also a serious risk to "the secrecy of ballots in any system that keeps paper records of individual ballots."[229]

Some jurisdictions, such as Venezuela, have long used fingerprint scanners to control entry to polling stations, and, as of 2012, thumb scans were also required to activate the electronic voting process.[230] This raised fears that how you voted was being recorded. This is creepy because the average person really has no idea what is going on inside a complex piece of electronic gear like a voting machine.

Nor do most people realize how many fingerprint scanners are used in industry. They are used by employees to clock in and out of work. Airport employees use a fingerprint scanner to access restricted areas, and travelers are increasingly being required to undergo finger or iris scans in the interest of security.

The biggest recent leap in the consumerization of fingerprint recognition was the 2013 introduction of Apple's iPhone 5S with its built-in fingerprint scanner. While Chaos Computer Club hackers defeated the technology within two days, biometrics will undoubtedly become commonplace on consumer devices.

If your phone does not have a fingerprint reader, don't fret: you can always download a novelty application like "Fingerprint Lock" for Android mobile phones. It does not actually scan your finger—it just pretends to do that, so you can impress your friends and give some illusion of security. You actually unlock your phone simply by touching a secret spot on the screen.

Like a $2.95 dummy security camera, this bit of "security theatrics" is probably better than no protection at all. It is also worth noting that Chinese-made Lenovo laptops have had fingerprint readers for almost a decade, and some businesses have opted to purchase this brand expressly for that reason.

One of the most interesting recent uses of fingerprint readers happened in an unlikely place, a chain of very high-end stores that sells things like Prada scarves and purses.

Every year the retailer hired temporary sales clerks for the busy pre-holiday sales period. This store, not wanting to annoy its elite clientele, also routinely accepted returns without requiring a purchase receipt. Often, in January, they found that they were unable to match a purchase transaction for the return of expensive items like $2,700 handbags. The culprits often turned out to be those temporary workers, who simply took the merchandise off the shelf and refunded it to their own credit cards.

To avoid detection, the dishonest salesperson would put the illegal return through on the number of an unwitting long-serving employee. Long after the temp had left the job, that veteran clerk would be hauled in to answer for a crime she never committed.

The retail chain's head of security came up with an ingenious and inexpensive solution: to this day you will notice sale clerks there discreetly reaching under the counter to access a fingerprint scanner which seals each and every transaction.

The main effect of the proliferation of fingerprint scanners is to ease us into the idea of routinely using our bodies as identification. Research is progressing on using other unique features of our bodies as biometric identification, from our "breath signature" to our cardiac rhythms. These days, the ones that read your finger usually check for a pulse, just to make sure the finger has not been borrowed.

Most people intuitively find death to be creepy, which is why almost every language has some euphemisms for it. Now, technology is finding ways to make it even creepier.

The Swedish company Pause Ljud & Bild is selling coffins outfitted with corpse-controllable music systems and 4G Internet access, just in case. The CataCoffin features "divine tweeters with external cooling and one hell of an eight-inch subwoofer, fine-tuned to the coffin's unique interior acoustic space." The developers have even gone to the trouble of arranging for a matching tombstone to supply subterranean power, and updatable playlists on Spotify.

At $30,000, these things are not exactly flying off the shelf. Most people seem more inclined to order the Lady de Guadalupe Steel Casket from Walmart.com for $1,199, though the concept of putting the remains of your loved one into a product ordered online from Walmart strikes some as an example of terminal cheapness.

No matter which coffin you choose, the South Koreans have created an interesting use for it. In a bizarre trend that speaks to the country's escalating suicide rate, South Koreans are turning to "fake funerals" to learn to appreciate their lives. Eulogies, final letters to loved ones, and time spent inside a coffin form part of the experience. *Vice Magazine* sent Yuka Uchida to have her own creepy "Well Dying" experience in the woods near Seoul.[231]

Uchida found the process useful in some capacity, suggesting that "by telling you that this is your final day, and making you focus on nothing but yourself, then making you enter the private space, the casket, this session creates an ideal opportunity for contemplation."

On a more modest scale, a company called LivesOn offers "your social afterlife." Those obsessed with Twitter can now depart this realm knowing that "when your heart stops beating, you'll keep tweeting." In a review of it, Theo Merz notes that the service, if it worked well, would allow us to get the three things he says everyone longs for: "1. To cheat death. 2. To see ourselves as others see us. 3. To have a second version of ourselves, which could deal with the drudgery of the everyday."[232]

The app uses artificial intelligence to learn about your style while you are still around, and then attempts to carry on your persona indefinitely. You may also appoint an Executor to manage your legacy account, and perhaps decide when you've finally tweeted enough.

LivesOn is currently a free service, and while it is still unclear how they will make any money, they do talk about a "Recommendation Engine" coming soon. Presumably your fine taste and good reputation during life, which you can no longer sully by any indiscretions, will be of some value after you're gone. Your favorite brand of headphones and those special romantic getaway hotels might continue to benefit from your glowing postmortem endorsements.

In planning for your life after death, you might also want to consider pet care. Various websites currently offer to care for pets after the Second Coming of Jesus Christ. One of these claims that "for a small donation of £69.99 pounds, we will make sure your pets are well fed and taken care of long after you and your family have been taken up."[233] Next to the picture of cute kittens and puppies, the site explains "Just because we are atheists doesn't mean we are not animal lovers. We adore all kinds of pets and would love to look after your pets after you are gone." However, they will take your money now. It might actually be cheaper to get your pet a mail order

Doctor of Divinity degree for $32.99 plus shipping and handling from TheMonastery.org.[234] That credential might carry some weight in a brutal post-Rapture world.

Techno-zealots like Ray Kurzweil assure us that we will soon be uploading our consciousness to a computer. In his 2005 book *The Singularity Is Near*, Kurzweil predicted that ongoing progress in biotechnology would mean that by the middle of the century, "humans will develop the means to instantly create new portions of ourselves, either biological or nonbiologicial," so that people can have "a biological body at one time and not at another, then have it again, then change it."[235]

He also claims there will soon be "software-based humans" who will "live out on the Web, projecting bodies whenever they need or want them, including holographically projected bodies, foglet-projected bodies and physical bodies comprising nanobot swarms."

There are certainly credible candidates for this already and we have met some of them, including Hank, the Coca-Cola chatterbot, and BINA48, the humanoid robot in Vermont. Perhaps the most compelling is IBM's Watson, which makes up in smarts and databases for what it lacks in physical body. IBM has created a whole new division to move Watson into real and profitable lines of business like health care and city planning.

Humanoid robots are creepy in several ways. If we don't know how the artificial consciousness is operating, it falls into the mysterious realm of technology. More frightening, perhaps, is the loss of control over the conversation when we find ourselves engaged in dialogue with what turns out to be a very intelligent non-human entity.

In this book, I've tried to be very forthcoming about technologies with the understanding that almost all can be used for good or for evil. There is one that gives me some ethical qualms, and that's the topic of "wireheading." I would just feel terrible if somebody used this book as a jumping-off point to engage in this practice and did harm to the most precious thing on earth, a human brain.

Still, there are hobbyists experimenting with direct brain stimulation in a quest for the ultimate pleasure, enthusiastically risking permanent brain damage. This quote, from the aptly named website highexistence.com, describes this nirvana they say they hope to attain:

> Imagine, for one second, the *ultimate* bliss. Ecstasy + that first kiss + the climax of your favorite song + winning the lottery + your best orgasm ever * 1,000,000,000. Pleasure so fantastic that is almost hurts. What if you could experience that 24/7 and <u>never</u> get bored with it?[236]

If the brain can be stimulated to heights of pleasure, can the reverse also happen? Since there is a dark side to every technology, interfering with the brain could be used for many nefarious purposes by torturers and other evil-doers.

In fact, government-sanctioned use of this type of brain manipulation is being discussed as a way to make those who commit particularly heinous crimes suffer for longer than is currently possible.

Writing on her blog, French philosopher Rebecca Roche mused that there are drugs that are known to alter our perception of the passing of time. Perhaps one could be used to trick the mind of a prisoner into thinking that he was serving, say, a one-thousand-year sentence. The same technology could be used by repressive regimes as a form of hideous torture.

Roche even suggests that the day may come when we can upload a prisoner's mind into a computer, and alter the mental cycle rate. "Uploading the mind of a convicted criminal and running it a million times faster than normal would enable the uploaded criminal to serve a one-thousand-year sentence in eight-and-a-half hours. This would, obviously, be much cheaper for the taxpayer than extending criminals' lifespans to enable them to serve 1,000 years in real time." She goes on to suggest that "the eight-and-a-half hour 1,000-year sentence could be followed by a few hours (or, from the point of view of the criminal, several hundred years) of treatment and rehabilitation."[237] So that vicious serial killer or hardened terrorist could be home in time for supper.

Time, it appears, is truly in the mind of the beholder.

# Time Creep

About a decade ago, I first heard the term "beware of time traveling robots from the future" from privacy advocate Brad Templeton, who's done great work with the Electronic Frontier Foundation and Singularity University. At the time, I didn't appreciate just how truly prescient it was.

As the cost of data storage has plummeted, it is cheaper and more convenient for most companies, and even individuals, to simply keep data around. Think about the last time you got a new computer. Chances are you "migrated" much if not all of the hard disk contents to a new, larger device, "just in case." It was easier than having to think about what to keep and what to throw away.

It is true that many companies have data retention polices and discard certain physical and electronic records on a scheduled basis. This certainly applies to financial files which have a retention period defined by tax authorities. The controversial "telephone metadata" files amassed by the NSA are supposed to be destroyed after five years, though the agency tried to fight that in court.[238] In some businesses, such as the brokerage industry, there are also rules on the retention of emails and chat logs to allow for investigation of possible insider trading.

Companies can also set voluntary data retention policies, and the good ones will share them with you. Google, for example, keeps your Web History indefinitely unless you have that feature disabled, in which case they are partially anonymized after eighteen months. As the Electronic Frontier Foundation points out, "Note that disabling Web History in your Google account will not prevent Google from gathering and storing this information and using it for internal purposes. It also does not change the fact that any information gathered and stored by Google could be sought by law enforcement."[239]

Most data falls into a gray area when it comes to retention. Your Word documents, your PowerPoint presentations, your blog postings, and those emails that you wrote and decided not to send fall outside of any formal policy.

When it comes to cloud storage, data is often automatically backed up on remote computers, possibly even in another country. Because of the far-reaching powers of the USA PATRIOT Act, there are Canadian organizations, including the Government of British Columbia, that forbid the storage of certain records outside of the province.

Sometimes, people are very glad that records are kept around. A Hamilton, Ontario resident lost a winning lottery ticket and was tracked down by officials of the Ontario Lottery and Gaming Commission (OLG) who handed her the $50 million prize.[240]

How did they find her? It turns out the OLG used a combination of credit card records and video surveillance footage from the drug store where she made the purchase.

OLG officials were reluctant to share their exact methodology with me for validating claims. They did, after all, receive 435 "inquiries" from people who were sure they had this winning ticket but lost it in the washing machine or in their dog's belly.

The OLG did tell me this much: it helped that there was only a single winning ticket, and they knew it was sold in Cambridge, Ontario on November 30, 2012. In addition to credit card records, they used interviews and video footage. They would not comment on how long stores like Shoppers Drug Mart retain their surveillance videos, but, given the timing in this case, it seems pretty likely that it's for a lot longer than thirty days.

In fact, with storage media becoming so inexpensive, it may well cost more in effort to delete digital videos than to leave them around indefinitely. In a 2007 report, Ontario's Information and Privacy Commissioner, Ann Cavoukian, recommended that video surveillance images in public places such as parks that have not been requested by law enforcement should be "routinely erased according to a standard schedule (normally between 48 and 72 hours)."[241]

The Internet's omnivorous storage capacity has made it a place where time can stand still, and even run backwards.

It's hard to imagine people getting nostalgic about the "olden days" of a technology such as the Internet that has only been around for a few decades. Yet people do wax poetic about "the classic Facebook" and even chat rooms like California's famous "Whole Earth 'Lectronic Link" (The WELL) which began in the 1980s as a dial-up bulletin board system.

On the WELL, people freely discussed hacking, The Grateful Dead, drugs, and virtual communities. It was co-founded by Stewart Brand whose "Whole Earth" catalogs kept city-bound 1960s hippies dreaming of becoming sheep farmers in New Zealand.

Anyone with Internet nostalgia can satisfy their longing by using the Wayback Machine (www.archive.org) to see what things were really like back then. You are likely to find your own or your employer's website from the 1990s and will probably grimace at what you see.

In a search to find some of the oldest discussion group posts, I trawled through USENET postings, an early platform used around the world. Consider this commentary from 1981 about the movie *Clash of the Titans* from the science fiction fan board.[242]

> Date: 17 Jun 1981 10:40:32-EDT
> From: cjh at CCA-UNIX (Chip Hitchcock)
> Subject: bad actors in CotT (CLoT?)
> I'm curious about your statement that the actress(?) playing Andromeda was so bad they had to hire a stand-in for the bath scene.
>
> Certainly her proportions were extreme enough to satisfy most people; was it that she refused to do a nude scene (which I find

thoroughly unlikely for an unknown in present-day filmmaking)? And do you think that one mark of a good actress is willingness to "strip for the camera?"

Chip's ancient remark, though it now seems rather sexist, shouldn't be too embarrassing to him. It certainly pales in front of some of the things public figures have said and done online. Hardly a week goes by without some politician having to apologize for a hasty tweet or ill-considered email. The smart ones are now letting their assistants run their accounts.

More and more conversations are being archived, whether you know it or not. Google's "chat" feature can easily be set up to capture both sides of your conversation and send it to Gmail. Any informal online exchange you had with somebody might be hauled out a few years from now to your detriment. Even taking your chat "off the record" is not enough, warns Google, since "if you're talking to someone who is connected to the network with a desktop chat client, it's possible that his or her software is keeping a separate copy of the chat history."[243]

Believe it or not, there was a time when you could take back an ill-considered email. In the 1980s, on internal corporate email systems like IBM's PROFS, messages sat on a computer disk until a recipient accessed it. If the mail was sent, but had not yet been read by the recipient, you could hit an "unsend" button and it was simply deleted from their inbox.

Many careers and relationships were probably saved by this feature.

Lt. Col. Oliver North and President Ronald Reagan probably wished they had an "unsend" button for some of their emails from the 1980s. Their exchanges came back to haunt them in the Iran-Contra Affair. Bill Gates also suffered from pesky email persistence. A January 5, 1996 electronic memo from him was introduced as evidence in Microsoft's antitrust trial.

Now, of course, your email instantly travels out of your control to various places on the Internet. The closest we have to an unsend button is the optional Google Labs "Undo Send" feature that gives you a precious few seconds of leeway to reconsider.

As a result of the revelations from Chelsea Manning in 2010 and Edward Snowden in 2013, we now also have incontrovertible evidence that we are indeed being watched electronically by our governments.

Despite NSA Director Keith Alexander's flag waving denials, which I witnessed firsthand at the Black Hat 2013 conference in Las Vegas, the evidence for abuses keeps getting more damning.[244] If the U.S. government wants information on you in digital form, they can almost certainly get it.

There's a real danger that the Manning and Snowden revelations may lead to the worst of two evils—a diminution in our national security coupled with the further erosion of our privacy as the people behind these programs find even sneakier ways to extract information.

In a 2013 policy paper I did for the Ottawa-based Strategic Studies Working Group, I carried out a fuller analysis of this challenge and made some recommendations for improving our information security landscape.[245]

They included:

- Don't rely on security through obscurity
- Minimize exposure to social engineering and related attacks through good policies and strict adherence
- Increase the granularity of information classification
- Use Defense in Depth to put multiple barriers in the way of an adversary
- Plan carefully for failure both of technology and of technology security using the "safe to fail" approach
- Apply machine learning, self-repairing technologies, bio- and neuro-morphic computing

- Institute continuous cybersecurity awareness programs
- Actively model threats through cybersecurity exercises, hackathons, and red team exercises

None of these can address the fundamental question of just how much, if at all, our governments should be spying on us. That is a political and social question that has to be answered outside this domain. However, once we have some social consensus on that fundamental issue, these recommendations can help in their effective implementation and possibly improve transparency.

Still, "I'm from the government and I'm here to help you" strikes fear into many hearts. Governments have special rights to collect data on many aspects of our lives. Now, it turns out, they are falling over each other to sell this data or even give it away to anyone who's interested.

# Government Creep

Governments around the world, including the U.S., the U.K., Canada, India, and even little Slovenia, are all urging citizens to do their business with their government in an online fashion. Advantages may include faster turnaround, greater accuracy, and freeing up government workers for the cases that truly need human attention, while others are handled by technology. In the course of doing this, governments automatically acquire a great deal of data.

In a *New York Times* piece, Brett Goldstein, former chief data and information officer for the city of Chicago, lays out a host of advantages to opening up government data vaults.[246] He notes that in San Francisco, sensors detect vacant parking spots and direct you to them, while adjusting parking rates in real time based on demand. Goldstein also describes how "speed limits can be algorithmically adjusted based on current road and traffic conditions to optimize for safe and efficient travel."

At an Open Data Summit in Montpelier, VT, "an employee from the (State) Agency of Agriculture alluded to the 'political cost' of pursuing open data, noting, for example, that some Vermonters balk at the idea of farm information being posted online."[247] It's always been poor form to ask a farmer how many acres he owns or a rancher how many head of cattle he runs, but now the curious may be able to make such queries electronically and anonymously.

New York City was a pioneer in the Open Data arena. In October 2009, with considerable fanfare, they released a number of municipal data sets. The next day, they had to take one down because it contained private email addresses as well as secret questions and answers. Despite their fast corrective action, once a dataset is exposed, even for a day, it is impossible to get it back. In the absence of specific logging and tracking procedures, it is often impossible to know precisely who has accessed or copied the information.

Government databases that contain personal information are particularly sensitive. One of the most interesting is the one containing the legally-required disclosure of political campaign contributions.

If you donate to a U.S. federal political campaign you are required to provide information such as your address and occupation which will make its way to the database of the Federal Elections Commission (FEC).

This is to conform to what the *Wall Street Journal*'s Laura Saunders called "a crazy quilt of federal tax and election laws."[248] She also explains that "donations by individuals of $200 or more in a year must be publicly disclosed to the FEC within as little as 20 days."

Rules differ for U.S. state and municipal elections, and, of course, in other countries. But the principle, that the public should know who's contributing to a particular candidate, is a pretty fundamental and universal one.

Peeking at political contribution data wasn't a big issue when you were required to make a specific written request to access this information. You often had to wait while bureaucrats handled your query manually; so only the highly motivated bothered to make a request.

Now, vast amounts of government data can be searched by anyone, anywhere, often anonymously. It can also be manipulated with the same data analysis tools that corporations and governments themselves use.

I settled on Philadelphia as the place to investigate the public disclosure of campaign contributions. Like New York, this city was making its government data broadly available. They were also attempting to provide free city-wide Wi-Fi Internet access. These records are online, so there was no need to physically go to Philadelphia. Everything you might want is right there at OpenDataPhilly.org.

Since this particular analysis was done several years ago, it is almost certain that some of the details about how the data is presented may have changed. However, unless the law changes, campaign finance data will always be available in some form.

It is important to understand the difference between accessing government data through a user query interface and downloading entire data sets. OpenDataPhilly allows both, at least for campaign finance data.

I plugged a particular name into the query interface. The system churned for several minutes (one of the disadvantages of using query apps as opposed to processing the data yourself) and it gave me my answer. Ronald Rivest of 41 A****** Street, Arlington, Massachusetts 02476 donated $200.00 on May 4, 2011 to Friends of Stephanie Singer. I have redacted part of the street name but the original database was completely forthcoming with the precise address.

By cross-checking this name and address with some other databases, a technique called "data jigsawing," I was able to say with reasonable confidence that this contributor was in fact a well-known MIT professor and researcher, who is in fact the "R" in the famous RSA Encryption algorithm.

I think I met him once at a conference, but I know Ron Rivest mainly by reputation. And, let me be clear that there is nothing wrong with someone who lives in Massachusetts donating to a local candidate in Philadelphia. Unusual, maybe, but certainly not illegal.

I knew to search for Rivest's name in this database after downloading and processing the data on my own personal computer, and graphing the geographic patterns of donations. A donation from faraway Arlington, MA, to a Philadelphia local candidate practically jumped off the page.

This geographical analysis revealed some other remarkable facts. It turns out a very significant number of people gave 1719 Spring Garden Street, Philadelphia, PA 19130 as their address on the contribution form.

Specifically, for one of the campaign finance data sets:

> 86 of 588 people with the surname Smith listed that address
> 63 of 426 people with the surname Johnson listed that address
> 77 of 400 people with the surname Williams listed that address.

A quick trip to Google Maps to see if this is the world's largest rooming house reveals that this is the location of the "Financial Office & Electrical Apprentice Training" operation of a Union, the IBEW Local 98.

It is easy to imagine some possible scenarios that would lead to so many supporters of certain candidates having a connection to the union office address, but that would take us into the realm of speculation. But why are addresses required in the first place? One plausible explanation is to differentiate people with similar names. There could be certainly more than one "J. Smith" in the campaign contributions file.

Johnson, Smith, and Williams are the most common surnames in America, so I went through the 1,414 people in the file with those surnames, looking for cases where there might be confusion.

Here's what I found:

> Among the 426 Johnsons there were three possibly ambiguous entries
> Among the 588 Smiths there were eight possibly ambiguous entries
> Among the 400 Williamses there were four possibly ambiguous entries

The question for policy makers is whether or not this small additional power of discrimination is worth the wholesale exposing of addresses. Many people, such as victims of domestic violence, have good and valid reasons to avoid having their address posted online for all to see. On the flip side, being able to plot donations on a geographic grid might reveal some interesting information about which parts of the city are supporting which candidate. In this case, however, being allowed to give a union address regardless of personal address appears to defeat the purpose.

Voting is another area where privacy seems vital. Yet each piece of new technology brings more disturbing elements to the ballot box. There's a controversy in some jurisdictions over whether or not it's OK to "tweet your vote" by taking a picture of your ballot at the polling place.

This came up in 2013 in Nova Scotia, when a political blogger posted a picture of his marked ballot on a social media website. He was told he might face a fine of up to $5,000 for violating the province's Elections Act. In the U.S., laws vary by locality but the smartest move is to keep your smartphone and digital camera in your pocket during the voting process.

I reviewed the 2010 Mayoral Election results in Edmonton, AB, in which one candidate won by a sizable margin. In fact, some of his challengers received zero votes in some of the voting stations, such as the "City Wide Hospital Vote." Let's hope that there was no one in the hospital at the time that he knows personally who claimed they voted for him, because the public record shows otherwise.

In general, when very small numbers are involved, good statistical reporting systems are supposed to report "not significant" instead of giving out information that would allow inferences about specific cases.

Of course, some candidates may be interested to know if they got three or seven or zero votes, so these results are still generally reported as exact numbers. However, in an election where thousands of votes are cast, voter privacy should probably trump candidate narcissism.

While making political contributions (and even voting, except in some places such as Australia) is optional, every property owner pays taxes. The associated database provides a powerful example of the balance between privacy and fairness in the world of governmental Open Data.

To ensure you are being treated fairly where taxes are based on property assessment, you need some information. What is your neighbor's assessment? What is it based on? Does the house next door have a fireplace? A two-car garage?

Historically, this information was made available in hard copy format. In Calgary, AB, for example, the property tax rolls were made available for about a month every year around the time when owners were allowed to appeal their annual assessments.

The tax data was printed out on paper and taken to community centers around town, where eagle-eyed staffers kept watch over them. You were free to make notes on what you saw, but not to take the book away or even photocopy it. Digital cameras and smartphones with cameras were not an issue back then.

Calgary now has a system which allows anyone to view anyone's tax assessment, at any time, 24/7, 365 days a year. Though a great improvement from the consumer's point of view, this system inevitably created problems.

Newspaper reporters went through the data to find the mansions with the highest property valuations and snapped photos of them. The owners of the "Priciest Houses in Town" objected to this attention, fearing kooks and burglars. They were told the data was a matter of public record.

Soon after the system, known as "Fairshare," was launched, everyone who cared and bothered to look knew the tax value (which mirrors market value in Calgary) of their ex-wife's home, their boss's mansion, and even their garbage collector's house. People grumbled about the invasion of their privacy, but the municipal government persevered.

Then, an unexpected "second-order" problem arose.

Companies who specialized in the business of tax appeals started to write detailed letters to individual taxpayers along the lines of "we see your property is assessed at X dollars. Did you know your next-door neighbors are at only Y and Z dollars? Hire us to get your assessment lowered."

Spooked by this new, derivative use of their information, the City of Calgary closed the system for a period of time, bringing it back in a much more privacy-friendly manner. You can still find the dollar figure for any property simply by keying in the address. You even see whole neighborhoods on a map. However, to get the juicier property data you need to create an account with a password. Your usage can be monitored, and is limited to a certain number of properties per day, in an attempt to thwart the wholesale downloading of data. The City also tracks the Internet address used to make queries and posted a legal disclaimer explicitly banning commercial use of the information.

The new system isn't perfect. Any good computer science student can probably find a way around the IP tracking and other limitations. The twenty inquiries per day limitation is also frustrating to people who may want to search all the properties in a complex for perfectly legitimate reasons. However, the new system has struck reasonable balance between making information available to those who need it, while keeping it from the merely curious and the actively malicious.

Governments are still struggling with the idea of "public information." The line is also blurring between governmental and privately gathered information, which are becoming mashed together into one big data cloud.

Certain life events open normally private matters more fully to public scrutiny, and a divorce is an excellent example of how privacy protection can slip in the interest of public information. Those detailed and sometimes embarrassing lists of assets, chattels, claims, and counterclaims will become a matter of public record in most jurisdictions.[249]

Anyone willing to head to the musty basement of a courthouse can usually find out who got the house, the car, even the family pet. A nine-year-old Dalmatian hound was the subject of a two million dollar divorce-related claim involving a Calgary millionaire and his soon-to-be ex-wife. The full settlement details were not made public but we do know that Dad got the dog and the press quoted his stepson as saying "my mom is very wealthy now."[250]

If records are digitized, they are probably available from a data broker like Little Rock, Arkansas-based Acxiom, which, according to the *New York Times*, has data on "about 500 million active consumers worldwide, with about 1,500 data points per person. That includes a majority of adults in the United States."[251]

Acxiom sucks in data and uses PersonicX, a data-based classification system worthy of the NSA. It pops you into one of seventy socioeconomic clusters and life stage groups with cute if somewhat judgmental names like "Mature Rustics," "Resolute Renters," and "Midtown Minivanners."[252] Acxiom also offers a "Race model."

The company can provide data on people who are smokers, gamblers, dieters, etc. They often glean this information using self-reported surveys on sites that offer promotional deals to people willing to complete them.

Completing a survey on one of these sites can sometimes be a humbling experience, as you get "screened out" for reasons of age, income, or some other unknown. But even if you didn't win a prize or earn some of their fake currency, the information you give up in these surveys is being filed, exchanged, sold, and used to target you whenever possible.

# Deception Creep

One of the creepiest aspects of technology is that you never really know who or what to believe anymore.

You open your email and see that a well-meaning friend has sent you an urgent message: a mutual acquaintance has been arrested for some unlikely transgression. Maybe it's a hoax; maybe not. It turns out there are numerous free websites that make it simple to generate credible looking stories that your friend or co-worker was arrested for "driving naked" or "having sex with a sheep."[253]

Even credible news sources get hacked, sometimes by intruders and sometimes by insiders. Long before there was an Internet, when the hard copy version of the *Encyclopedia Britannica* was the world's definitive reference work, I heard about an insider there who used a text editor on the encyclopedia's word processing files to substitute "the Prophet Muhammad" for every instance of "Jesus Christ." The giant, expensive book set nearly went to print that way.

Online pranksters such as "The Yes Men" have also wreaked havoc. In one case, they were able to tank the price of the shares of Union Carbide by putting out bogus news stories about the Bhopal disaster and how the company was planning to atone for it.

The proliferation of images on the Internet has taken hoaxing to a whole new level. There is ample photographic evidence of Barack Obama smoking all kinds of things all over the Internet. Most of it is poorly fabricated. By doing an "image search" on Google Images or TinEye we can even track all the places a particular photo has been posted. I did this during the first Obama campaign and found that an apparently genuine image of him smoking a cigarette had even made its way onto Stormfront.com, a white supremacist site that has been called "the Internet's first hate site."[254]

I once asked a photo editor at the *New York Times* if he could detect faked photographs now that they could be digitally manipulated. He said that in the past, he would look for things like sharp edges where negatives had been cut and pasted together. Now that image tampering is done by computers, he acknowledged that fakes can certainly get past him.

Ordinary citizens are even starting to practice various forms of disinformation. Shopper Keith Gormezano posted his Safeway Club Card online, encouraging others to download the barcode and use it when they went shopping. It worked, and now there are people all over North American hearing "thank you, Mr. Gormezano" from grocery store cashiers. As one blogger pointed out, this ruse might someday backfire on Mr. Gormezano if authorities try to figure out why he consumed 666 bottles of wine last month.

Nowhere is consumer lying more blatant than in polls and on surveys. The most recent provincial election polls in Western Canada have all been wildly different from the actual voting results. In 2013 Angus Reid and Ekos predicted a crushing defeat for Premier Christy Clark in British Columbia; instead she was easily re-elected.

Do voters simply wake up on Election Day and switch allegiances? Telephone polls usually focus on people who have landline telephones. So the pollsters probably heard from a lot of octogenarians, missing the entirety of the youth vote. And, of course, people have been known to lie, telling pollsters what they think they want to hear, or whatever will get them off the phone quickly.

Online consumer surveys sometimes seek to compensate you for your time with cash incentives or points that can be converted into something like a magazine subscription. The elderly are usually screened from these surveys, reflecting the way the market views them, generally speaking.

Cut your age down to thirty-nine, and the surveys will flow like water. After completing one, for fun and points, on a medical condition I don't have (hemorrhoids), one of these surveys gleefully asked me probing questions about what I have around the house in a bizarre attempt to profile me. Do I have a bike? A Yoga mat? Cast iron skillet? Face paint? Scented candles? A cape?

The same survey asked me to select the person I'd most like to take for a Sunday walk. I struggled to choose among the options presented: Warren Buffet, Jesus, or Tina Fey.

My favorite story about trusting technology over common sense comes from the early days of product scanners and point-of-sale (POS) terminals at the supermarket. Back then the checkout clerks were required to call out the purchases and prices verbally—to reassure shoppers that this technology was really working.

Some innovative hackers taped over bar codes of various products, substituting bar codes from more expensive items so they wouldn't really be stealing:

Store clerk: "Lobster tail: $14.99"

Customer (pointing to bag of potatoes): "No, that's a sack of potatoes"

Store clerk (pointing to POS terminal screen): "Lobster tail, $14.99."

Watching a human being allow a computer readout to take precedence over what she saw with her own eyes was both funny and profoundly unsettling.

In the 1980s and 1990s, radio station contests were sometimes rigged so the station manager's cousin would win the trip by being, remarkably, the fifth caller and our lucky winner." Hackers like Kevin Poulsen figured out ways to attack a broadcaster's phone system to win some high stakes contests.

As reported by Alan Wlasuk on TechRepublic, Poulsen's "iconic 1991 hack was a takeover of all of the telephone lines for the Los Angeles KIIS-FM radio station, guaranteeing that he would be the 102nd caller and win the prize of a Porsche 944 S2. The bold Poulsen was wanted by the FBI for federal computer hacking at the same time he was winning the Porsche and $20,000 in prize money at a separate station. Poulsen spent 51 months in a federal prison, the longest sentence of a cybercriminal at that time."[255]

The Internet has allowed even more creative tampering with contests. An unsanctioned "Who Wants Justin the Most?" competition in 2010 claimed that Justin Bieber would perform in whichever country tallied the most fan votes. Internet pranksters rigged the contest so North Korea won, with 659,141 votes. Bieber did not go to North Korea.[256]

In a similar fiasco, a 2013 "Meet Taylor Swift" contest was canceled after the Boston radio station that sponsored it reported that "the integrity of the 'Taylor Swift's Biggest Fan' contest" had been "compromised." Reddit and 4chan promoted a thirty-nine-year-old bearded man knows as "Charles Z." "Crush the dreams of these girls," urged 4chan, "and give him a chance to make a complete ass of himself by blatantly just sniffing her hair with cameras rolling."[257]

In 2008, the Canadian Broadcasting Corporation ran a contest for a new composition to replace its classic "Hockey Night in Canada" theme song. Vying for the $100,000 first prize, young Logan Aube submitted an entry that he described as "mostly comprised of cat and sheep sounds, baby cries, and gunshots/explosions."

Using the power of the Internet, specifically the Something Awful forum, YouTube, and Facebook, Aube managed to push his entry into first place. The CBC faced angry protests when they took it down.[258]

Another malicious activity happening online with alarming frequency is the hacking of email accounts. According to a report in *The Economist*, "One day in early 2010, an American working for an environmental NGO in China noticed something odd happening to his BlackBerry; it was sending an email from his account without his doing."[259]

He watched, dumbfounded, as the email went out to a long list of U.S. government recipients, none of which was in his address book. Seconds later he saw the email disappear from his sent folder.

Eventually he heard from the FBI that his email account and those of several colleagues had been compromised by hackers from China. All the victims had attended a climate-change conference in Copenhagen in December 2009, where America and China had clashed.

David Barboza, a journalist at the *New York Times*, reported in October 2012 that relatives of Wen Jiabao, China's prime minister at the time, had huge fortunes. After his story was published, Chinese hackers compromised the publication's networks to get at Mr. Barboza's work email account. "Other news organizations, including the *Wall Street Journal* and Reuters, noticed similar Chinese intrusions."[260]

Even hardware is no longer safe.

At the 2012 Black Hat conference in Las Vegas, Jonathan Brossard gave the world a peek into the secret world of hardware back doors, which are a lot harder to detect than software ones, and virtually impossible to fix once they are installed.[261]

Brossard fired up a normal-looking computer with a diddled BIOS chip, the software that controls how a computer starts up. This was enough to disable security features of Microsoft's latest Windows 7 operating system. In fact, it could have disabled any operating system, because it bypassed low-level security instructions in the computer's CPU. He made the additional point that much of this nasty exploit is "built on top of free software, including the Coreboot project, meaning that most of its source code is already public." So, unlike hacks that require microscopes and cutting chips apart, this one is done with easy-to-obtain tools and some brainpower. It is also safely beyond the reach of antivirus software: even erasing the hard disk and reloading the operating system won't do a thing to it.[262]

The clear implication is that if someone can obtain physical access to a computer, especially at the manufacturer or distributor level, they can "own" it forever, making it take instructions from them over the Internet at will.

While an Intel spokesperson shot back that this was largely a theoretical vulnerability, there is certainly evidence of hardware back doors such as the Stuxnet worm that have been much more than theoretical.

From the earliest days of computers, people tried to make them do unlikely and sometimes humorous things. In the days when computers were the size of rooms, we would sometimes prank the operators by making their huge clunky line printers play musical sounds. Another favorite was EDITH, which displayed an image of a naked woman, made entirely of ASCII characters like Xs and *s on the printer.

Often programmers built in secret instructions to display their names as the authors, or for their own convenience while testing or using the program in the future. We called them features, but our bosses viewed them as unacceptable holes in the system. So we made sure they never found out. Nobody was ever going to read our code line by line anyway, except perhaps another programmer, who would then be in on the little secret.

In the 1970s I worked on the MULTICS operating system, which was explicitly and carefully designed for security. An ancient (1974) report on this system contains these prophetic words: "the penetrator can install 'trap doors' in the system which permit him access, but are virtually undetectable."[263]

By the time the 1983 film *WarGames* was made, the term "backdoor" was *en vogue*. In that movie, the secret access code was the name of a character's dead son, Joshua.

In four decades of watching hackers, I've come to admire both their ingenuity and their persistence. The best way to summarize the goal of the hacker mind, at least for the "White Hat ones," is "not to do what you're not supposed to do—to do what you're not supposed to be able to do."

System designers often fail to "expect the unexpected." One of my favorite examples is a German author who embedded SQL commands into a book he published. Exploiting a flaw in the Amazon web store, he arranged it so people who tried to "Look Inside" his book had their browser redirected to the page for purchasing it. Amazon quickly fixed the problem but it was one of the more clever attacks of this nature.[264]

Unintentional "magic strings" can also happen. In September 2013, word started to spread about a "magic sequence" of Arabic letters that would crash iPhones and other devices that used Apple's CoreText text rendering system. According to *Business Insider*, "just to read these letters in your timeline was enough to crash your Twitter app."[265]

Are those who exploit such weaknesses good or evil? Ultimately that depends on their motivation and how they use the knowledge that is hidden from the masses. One thing is for certain: technology tricks can be used to harm us, often in the pocketbook.

Most people know that there's no terminally ill rich widow in Nigeria waiting to share her fortune with a lovely person like you, and that Bill Gates doesn't randomly select email addresses to send out million dollar checks. "Get your free credit report" often means "give us enough information to steal your identity." Even "I want to buy that guitar you listed for sale on Kijiji" could actually translate into "I'm harvesting and selling confirmed email addresses and want to add yours to the list."

Scam artists are using some downright insidious tricks to tug at our heartstrings. According to a report in *The Guardian*, "Peter Saunders from Edinburgh received a heart-wrenching letter from Namukula Viola of Uganda. She is just 16, with two younger sisters, orphaned when her mother was raped and killed by rebels in fighting in the north of the country."[266]

He was horrified by her detailed tale of woe, and sent money as requested. Then a news report alerted Saunders to the fact that a lot of people were getting identical sob story letters. It was in fact a wicked hoax aimed at extracting money from nice but gullible people. An interesting twist is that many of the victims were artists, probably because their names and addresses were listed in a certain directory.

Even the savviest users can get fooled by a type of online trickery called Dark Patterns. As explained on darkpatterns.org, this is "a type of user interface that appears to have been carefully crafted to trick users into doing things, such as buying insurance with their purchase or signing up for recurring bills."

A wonderful example appeared on the website of Ryanair, which offers inexpensive air tickets but charges for just about everything else like checked baggage and boarding pass printing. One of the things they sell is travel insurance. At one point their site was pushing it so hard that you had to dig for the "opt out" option hidden between Latvia and Lithuania. It's not even in alphabetical order! If you didn't spot the "No Travel Insurance Required" choice and specifically select it you were continuously taken back to "Please select country of residence." Writing on DarkPatterns.org, Harry Brignull observed that "What's interesting about this pattern is that it gives the site owner plausible deniability: they can claim that when you read the words on the page, it's entirely clear what's being said, so what's the problem?"[267] You can judge this one for yourself at http://darkpatterns.org/library/trick_questions/.

Even more people have been struck by the Conduit search bar, a piece of software that frequently shows up after someone has downloaded "free software from a reputable site." These free software download sites need to make money, so their "automatic" installation often brings programs like Conduit along for the ride.

The website malwaretips.com says this about Conduit: "it's technically not a virus, but it does exhibit plenty of malicious traits, such as rootkit capabilities to hook deep into the operating system, browser hijacking, and in general just interfering with the user experience. The industry generally refers to it as a 'PUP,' or potentially unwanted program."[268] To make matters worse, some of the "free Conduit removal tools" offered online to desperate victims are themselves vicious pieces of malware.

Even if you aren't infected with malware, you automatically give out information every time you use the Internet. Ever notice how the people who "lost 40 pounds using this one weird trick" or "are waiting to meet you" magically seem to live in your hometown? That's because the ads are being customized to your physical location.

But how do they know where you live? The simple answer is that your Internet Protocol (IP) address is generally tied to the geographical location of your Internet Service Provider (ISP). For the Internet to work properly, your return address has to be sent out with each packet transmitted. IP localization services like http://www.geobytes.com/iplocator.htm not only give the country code for pretty much any IP address, they even return a latitude and longitude. If you work for a large enough company, it may be serving as your ISP, so your IP address might effectively reveal your employer.

I gained an appreciation for just how far off from reality an IP address can be when the Canadian Forces invited me to visit their operations in Afghanistan. We stopped enroute at Camp Mirage, a now-closed military airfield that we were told to describe as "somewhere in Southwest Asia."

We had a pretty good idea what country we were in, but to check, I logged on from one of the courtesy Internet terminals and asked a tech savvy friend in Canada to attempt to geo-locate me. He came back with an address in Ottawa. This made sense since that's probably where our military Internet traffic was actually entering the public Internet. But in terms of actually locating me, the IP address location was off by over 10,000 kilometers.

I've had similar experiences working on forensic investigations where the location of the address we're trying to trace often comes up as a facility of the Internet Service Provider.

Many Canadians complain that they have a more restrictive version of Netflix and no access to a lot of things that tech savvy Americans take for granted like Hulu Plus, Pandora, and Oyster. These services provide access to on-demand media (videos, music, and books respectively) but are geo-fenced to U.S. users only because of licensing restrictions.

Creative Canadians have devised and published methods for faking a physical presence in the U.S. Often they involve using a Virtual Private Network (VPN) service to make your traffic appear to originate from the country of your choice. You can find out about all this and more at sites like www.howtogetitincanada.com.

Tweaking your position in cyberspace is an interesting and popular hack. But being deceived about your real-world location can have serious consequences. At a technical security seminar in 2007, researchers from an Italian information technology company gave a truly creepy demonstration called "How to freak out your Satellite Navigation." Starting with a stock vehicle, they showed how to hack RDS-TMC (Radio Data System–Traffic Message Channel), the FM Radio system that provides traffic data to car navigation systems.[269]

Using "a PC and some cheap home made electronics," they were able to inject messages into the Honda's navigation system ranging from "Traffic Queueing" to "Bomb Alert" to the ever-popular "Bull Fight." More menacingly, if they marked a road, bridge, or tunnel as "Code 401–Road Closed," the system would silently plan and suggest another route. Being able to control, or at least seriously influence, somebody's driving behavior at a distance could be a terrorist's dream scenario.

Even if you're not a slave to online driving directions, you probably rely on other digital guides for directions. Even Siri, the trusted voice of Apple's personal assistant, can lead you astray, and she is certainly keeping track of you. According to Nicole Ozer of the American Civil Liberties Union of Northern California, Siri stores a trove of personal information including the people in your contact list, your music preferences, and even how you label your email.[270] Ozer notes that "This data can be really personal, like if you ask Siri, 'where is the nearest abortion clinic?'." She adds that Apple reserves the right to share your data with "Apple's partners who are providing related services to Apple."

Not only can Siri spy on you, she might even misdirect you or hold back information. As the ACLU noted in 2011, Siri came up blank on "birth control information" and was instead directing people seeking abortions to pregnancy crisis centers that discourage abortions.[271]

There may be a pretty straightforward, non-malicious explanation for this: abortion clinics rarely use the word "abortion" in their name or listings. And to be fair, artificial intelligence is definitely improving how personal digital assistants function. However, people should still be concerned about secrets being exchanged behind their backs by systems to which they do not have direct access.

Humans, of course, have their own secret signals. For years, New York City "meter maids," now called "parking agents," would put a bag of M&M's candy on their own dashboards, thereby fending off tickets from their fellow agents. It turns out that "secret handshakes" like this are all over the place, especially in electronic technology where they often rest undiscovered until somebody stumbles upon them. Their functionality is always there, but hidden in plain sight, available only to the initiated.

If you ever see somebody in a BMW doing this ritual:

1. Get in and close all doors.
2. Turn on the ignition and turn off quickly. (No more than five seconds) to start the process. Next action must take place within thirty seconds.
3. Remove the 1st key.
4. Hold the key up near your left shoulder (this is so it is closer to the remote receiver antenna.
5. Hold down the unlock button and press the lock button three times. Release the unlock button and the doors lock which confirms the operation.[272]

They're probably not trying to steal the car. Instead, they are following a semi-secret procedure designed by the car's manufacturer to program an ignition "chip key."

There are all sorts of codes lurking inside cars, especially luxury models. Some can even unlock a vehicle and start its engine. If you have a microscope, a supply of valid keys to cut apart, and a lot of patience, you can discover how this works through a technique called "chip slicing."

However, revealing those secrets can get you into trouble. Flavio Garcia, a lecturer in computer science at the University of Birmingham, planned to present a paper called "Dismantling Megamos Crypto: Wirelessly Lockpicking a Vehicle Immobiliser" at an academic conference. He promised to divulge the secret codes of luxury automobiles "including Porsches, Audis, Bentleys and Lamborghinis."

Instead, he discovered that the U.K. High Court doesn't take lightly to hacking the types of cars driven by the wealthy, such as judges and lawyers. Garcia was slapped with an injunction and prohibited from publishing his findings. However, the odds are very good that those very codes are out there on the Internet, if you know where to look.

Hoaxes and deception are everywhere on the Internet, along with the tools to perpetrate them. Even Google is not immune. Although the company has started taking itself much more seriously, Google still stages an annual April Fool's Day hoax and allows its staff to plant hidden features called "Easter Eggs" in some of its software.

As explained on the *mental_floss* website, there are various hidden Easter Eggs, jokes, and timewasters in almost every Google service, product, or new device."[273]

Here is one you can try:

1. Go to YouTube
2. Start watching a video
3. Click outside the search bar
4. Type "1980"

"This will launch a playable game of Missile Command above the video. Beware! The aliens are trying to destroy the video you're watching."

Hoaxes are fun, but deception can be both effective and lucrative.

The news site reddit has acknowledged that, in that site's early days, they spawned a fleet of fake accounts, often creating a new user every time they made an entry.

As Derek Mead wrote on Motherboard.com, "by populating the site with accounts whose strings they pulled, the reddit crew could shape the discourse and sharing of the site in the direction they wanted, and as the real user base grew, those standards held, allowing the fake accounts to fade away."[274]

Online scammers have impersonated charities, victims of diseases, and even the FBI. Many of the scammers are in African countries, and the phenomenon is collectively referred to as a "419 fraud" in honor of a section of the Nigerian Criminal Code which seems to be rather laxly enforced.

While not a recommended hobby, some people do engage in conversations with the scammers, telling them wild stories, leading them on, and even asking for photographs of their passports.[275]

Still, it's best to simply delete those "too good to be true" emails, not open attachments, and spurn unsolicited online proposals no matter how attractive. Just walk away, so to speak.

Of course, that's difficult if someone is pointing a gun at you.

# Physible Creep

Handguns have been around since the 16th century, and their core technology hasn't really changed much. There has been some recent progress in building "smart guns" which use biometrics to respond only to the registered owner's voice or body. Perhaps the day is coming when guns can be fired by mere mental commands. All this enhanced security may be moot if an eight-year-old with a 3D printer and some plastic can run off a working handgun and take it to school.

Texas law student Cody Wilson and his non-profit corporation Defense Distributed caused a furor in 2013 when they "released the files for the Liberator pistol—the culmination of the Wiki Weapon Project."[276] Lawmakers at all levels launched into action trying to ban them. In November 2013, Philadelphia had the distinction of becoming the first city to outlaw the manufacturing of guns by 3D printers.[277] Critics quickly pointed out loopholes in that legislation. For example, it only bans the manufacture of 3D firearms in that city. Nothing in it makes it illegal to possess, say, a 3D printed gun created in neighboring Trenton, New Jersey. Further laws will close this loophole; then others will surely appear.

3D printed plastic guns are a high profile example of a new category of things called "Physibles." The Pirate Bay, which houses a repository of physibles, defines them as "data objects that are able (and feasible) to become physical."[278] 3D guns are perfect crime weapons. At around $25 for some plastic plus the use of a printer, they are disposable like "burner cell phones." Anyone who watches TV crime shows knows that police do their ballistic analysis of weapons by matching up the marks left on bullets. A gun that is used once and then discarded is fundamentally immune to that kind of forensics.

Most 3D printed guns have no serial numbers or identification marks and are made of plastic except for the firing pin, which in the case of The Liberator is an ordinary nail that you can buy on the Defense Distributed website for $5 with free shipping. So they are likely to pass through metal detectors, at least if you remove the nail, which Americans should not do because of a law called the Undetectable Firearms Act, which the U.S. Congress has extended until 2023.

Trying to engineer 3D printers to disallow the printing of guns is ultimately futile. It would simply lead to more innovative 3D guns that look like shower rings or action figures, two of the things commonly run off on 3D printers.

The attempt to control 3D printed guns demonstrates the emerging complexity of trying to separate the virtual and physical worlds. The CAD instruction files that allow the creation of the gun are clearly virtual. They have all the characteristics of digital information, such as instant accessibility all over the world, infinite replicability, and the inability to destroy them once they are distributed. Once they have been used to make a gun, they are transformed into a physical object with all its normal properties. The platform technology (3D printers) to make this happen is widely available, useful for many other functions, and almost impossible to control.

Lest you look at a plastic 3D gun and decide it would probably explode in your hand, please know that the 3D printing of metal is becoming a reality, through a process called direct metal laser sintering. Scanners are also getting better and cheaper. What comes out of 3D printers now will look like toys in a few years.

If 3D guns have law enforcement worried, other uses of 3D printing have some manufacturers terrified. The crime they have in mind is not murder, but the theft of intellectual property. A group at Michigan Technical University (MTU) went to the Thingiverse open source repository of 3D instruction files and selected, from over 100,000 items available, twenty that might be useful around the house and that could be bought commercially. Toys, watchbands, iPhone holders, pierogi makers, an orthotic insole, and the ever-popular shower curtain rings. Those are often the poster child for 3D printing because if you break one, you can scan one of its mates, 3D print it, and save yourself from buying a whole new set. As 3D printers get faster, it may even be quicker to run off a new garlic press than to rummage around looking for your old one.

The people who analyzed the 100,000 3D designs found that "even making the extremely conservative assumption that the household would only use the printer to make the selected twenty products a year, the avoided purchase cost savings would range from about $300 to $2000/year."[279]

That means the printer could pay for itself in anywhere from four months to two years, even counting materials costs. They used a printer called the RepRap, about half of whose parts can themselves be 3D printed in an eerie kind of self-replicating robot printer universe.

By doing for manufacturing what online banking did for the financial industry, mass 3D printing will essentially reverse the "one size fits all model" that has dominated manufacturing since Henry Ford built his assembly line. In fact, the MTU researchers found that "the largest savings (e.g. over 10,000%) were seen with individually customized products, such as the orthotic" and indeed that's where many see the future going. Rather than a trip to the shoe store, or even a visit to Amazon.com, your kids will select their colors and designs and run shoes off in time to wear to school tomorrow.

Criminals have already figured out that they can 3D print the keys to many cars, homes, and offices. A German lockpicking enthusiast apparently ran off a set of keys to unlock handcuffs from a photograph of the key hanging on a police officer's belt. They also find the printers great for running off skimmers to put over ATM slots.[280]

Even famous works of art are falling prey to scanning and 3D printing. Todd Blatt walked into the Walters Art Museum in Baltimore and used his Google Glass to take a sequence of thirty photographs of a bust of Marcus Aurelius.[281] He then stitched them together and sent them to a 3D printer. The result, as posted on his blog, is not going to pass for the original, but it is a reasonable memento of a trip to the art museum.[282]

In a paper on the legal aspects of 3D printing, Michael Weinberg argues that "the line between a physical object and a digital description of a physical object may also begin to blur. With a 3D printer, having the bits is almost as good as having the atoms."[283] He predicts that the kind of measures used to try to counter software, video, and music piracy will soon be extended to 3D designs.

There are some very creepy things possible with 3D printers, and some significant social changes. Handing your car keys to a parking valet may be risky business; he might scan them and 3D print himself a set. Then again, the days of car keys may be over as we move to keyfobs and biometrically controlled vehicles.

Because of the economics of factories and mass production, we have come to assume that if we see a ballpoint pen, it is actually a ballpoint pen. Pens that are also tear gas guns or video-recorders are usually sold in specialized "spy shops" and cost a lot more. Now, within reason, anything can be anything else.

There are even some bleeding edge experiments to alter the fundamental properties of matter. At the 2013 annual meeting of the American Academy for the Advancement of Sciences, Professor Anne Glover, CBE, and Chief Scientific Advisor to the President of the European Commission, thumped the table and noted that we now understand that the table is "99.9% nothing" yet "I could spend my entire life pounding on it and my hand would not go through it."

She added that it weighed tens of kilos and it would take a lot of energy to move it. "But, with the kind of work they are doing at

CERN," (the European research lab that gave us, among many other things, the world wide web) "we might be able to change that (its effective weight) to a few grams."[284]

With that tantalizing tidbit, she opened up a world where materials would behave in totally different ways from what we expect.

We are already beginning to move into the exciting realm of printing living things, or at least replacement parts for them. Just as Dolly became the poster sheep for mammalian cloning from adult body cells, Buttercup is the poster duck for 3D printing of appendages. Born with a backwards left foot, he recently received a 3D printed prosthetic copy of his sister's foot.

The foot is made of silicone, and scientists are now hard at work to use 3D bioprinters to make precisely fitting bone pieces. Researchers at the University of Nottingham predict that tissue replacement technology will be deployed and clinically approved within the next decade.[285]

Imagine going into minor surgery for nasal polyps and waking up with half your face removed. That happened to Eric Moger from Waltham Abbey, Essex, when surgeons discovered an aggressive facial tumor. Through computer design and 3D printing, Moger has received a 3D printed prosthetic face that has not only improved his appearance, it allows him to drink and eat normally instead of through a feeding tube.[286]

This technology has even found its way into science fairs. A seventeen-year-old American at the White House Science Fair showed off a $250 prosthetic hand printed on a 3D printer. The possibility of printing entire organs, and perhaps even bioengineered critters, is certainly looming. Experts predict that skin will probably be the first organ that can be printed in this fashion.[287] NASA has funded a project to use a 3D printer loaded with cartridges of powders and oil with a thirty -year shelf life to make things like pizzas in space.[288]

The combination of 3D printing technology with advances in biology, especially areas like stem cell research, may eventually take us to a world where if "we can dream it, we can make it." In fact, the New York-based company MakerLove.com is offering free downloadable files for printing your own sex toys. They say they have "been helping people avoid embarrassment" since 1998. Not content with the usual, you can choose to have your vibrator personalized with a 3D image of Sigmund Freud.

The company ran a competition among its fans to determine whose head should go on their next product. Guess who won? As explained on their website: "I don't believe everyone actually wants a sex toy made to look like Justin Bieber; in fact I think most people wanted to parody him in the form of one."

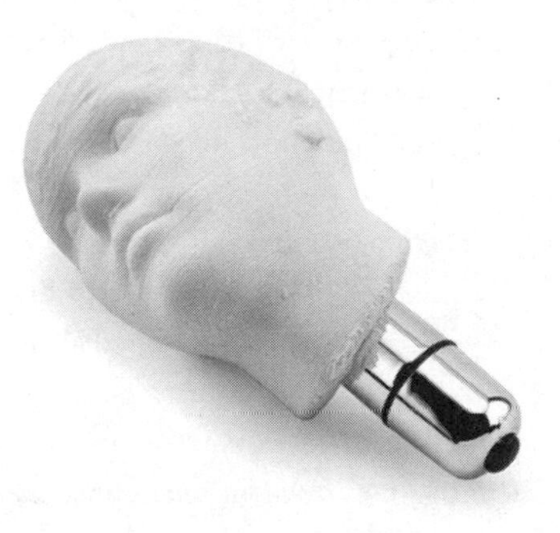

Figure 8. Justin Bieber 3D printed sex toy. Courtesy of Tom Nardone.

The New York Toy Collective takes sexy 3D printing to an even more personal level: "We are the first and only company to use 3D scanning technology to allow consumers to create sex toys modeled after their own bodies. We can scan a penis (or any part) and turn it into a silicone toy."[289] They even promise to preserve your unique vein structure on your cloned genitalia.

There's no doubt that 3D printing will revolutionize many aspects of our lives. Still, even in the future, we will probably continue to create our offspring in the time-honored way. It's what will happen to them from the moment of conception that's going to be both fascinating and very creepy.

# Child Creep

According to BabyCenter.com, "one in three children born in the United States already have an online presence before they are born. That number grows to 92 percent by the time they are two. In 2012 the average digital birth of children occurs at approximately six months."[290]

Mothers-to-be are often offered a DVD of their ultrasound examination, which presumably qualifies as baby's first photo shoot. A Japanese clinic has flirted with also sending them home with a 3D printed model of their developing fetus, called "The Shape of an Angel." It's produced by a special technology developed by Fasotec Company Ltd. of Japan, now part of Stratasys. The resin effigy of your unborn child, costing about $1,300, is delivered in an ornate jewelry box.[291]

When I put the question "what's the creepiest use of technology" to New York-based computer security expert Raj Goel, he didn't hesitate in answering: "What we're doing to kids."

He wasn't only thinking of child pornography and online sexual exploitation, but also the whole spectrum of images of children that we post online. There is a whole economy surrounding photographs and personal information, and it is migrating from corporate America into junior high schools.

LinkedIn is the dominant site for business networking and trading mutually beneficial contacts and endorsements. In August 2013, they announced they were changing their policy to allow kids as young as thirteen to have accounts:

> "We are updating our User Agreement to make LinkedIn available to students 13 years and older. Smart, ambitious students are already thinking about their futures when they step foot into high school—where they want to go to college, what they want to study,

where they want to live and work. We want to encourage these students to leverage the insights and connections of the millions of successful professionals on LinkedIn, so they can make the most informed decisions and start their careers off right."[292]

Technology blogger Graham Cluley called this move "shameful and creepy" and pointed out that "they missed out the bit about how LinkedIn also wants to boost its membership numbers, and offer a larger audience for its advertisers."[293] Proof of this is the simultaneous announcement of "University pages"—an opportunity for postsecondary institutes to reach out to prospective customers.

Other critics of LinkedIn for Kids noted that social media pressure and bullying have led to numerous youth suicides, and opined that the last thing they need is another social media pressure cooker, especially one related to future career prospects.

Jack Rivlin, writing in *The Telegraph,* also chastised LinkedIn in a story headlined "it's official, we've ruined childhood."[294] Rivlin sagely suggests that "LinkedIn, with its unabashed celebration of self-serving networking and creepy ability to seek out people you don't want to speak to, is exactly what we should be insulating kids from."

Rivlin notes that "LinkedIn tells you every decision about your future must be made in reference to 'marketable skills' and 'building a personal brand'." It's also worth mentioning that Jump Rope and Coloring are now skills you can list on your profile.

To be fair, LinkedIn is making some special privacy provisions for users who (self-report) their age as being less than eighteen. No company wants to run afoul of the Children's Online Privacy Protection Act (COPPA). However, plenty of parents would probably agree to their child's LinkedIn profile being viewed by Satan if it would help them get into Harvard or score a prestigious internship in the future.

Not wanting to be left behind, Facebook's manager of privacy and safety Nicky Jackson Colaco said in an interview in October 2013 that her company has "thought a lot about" lowering the minimum age of

Facebook users, which is currently also thirteen. There's a lot of purchasing influence, if not actual credit cards, in the hands of the pre-teen set.[295]

If the economic clout of children was ever in doubt, just consider the "Most Magical Place on Earth." In 2013, Walt Disney World in Orlando introduced MagicBands: durable, adjustable vinyl bracelets made to fit both children and adults. The bands are used for park admission, for unlocking hotel doors at Disney properties, and for goods and services on the grounds.

They will allow your kids to make certain purchases, helping even the youngest get into the habit of putting things on Mom and Dad's tab. Disney is even selling Minnie and Mickey MagicSliders, little charms to decorate the MagicBands and help children determine specifically where the RFID chip is located, to make it easier for them to "tap to play."

Buried in the fine print of the Disney Park Experience Terms and Conditions is a note that hotel "guests ages ten and over on your reservation will automatically be given charging privileges." Of course you can go to the front desk and get that revoked. But you'd better not ditch the wristbands entirely, because, aside from being your entry pass, they're the key that unlocks Disney's popular FastPass+ system with its line jumping privileges.

Most people probably assume that the technology in the bands is passive RFID, the kind of chip in credit cards and key fobs that allows you to pay for gasoline, and which are hidden inside purchases that you might be tempted to shoplift, like those infamous Mach 3 razor blades.

Actually, according to the Disney site, "each MagicBand contains an HF Radio Frequency device and a transmitter which sends and receives RF signals through a small antenna inside the MagicBand and enables it to be detected at short-range touch points throughout Walt Disney World Resort. MagicBands can also be read by long-range readers located at Walt Disney World Resort used to deliver personalized experiences, as well as provide information that helps us improve the overall experience in our parks."

For those who would prefer not to have the system alert Goofy or Snow White of their impending arrival, the card version of the MagicBand has "a passive HF Radio Frequency chip and cannot be detected by the long-range readers."

There are secret technologies all over a Disney park, and even people hidden just below the surface. Some commentators have observed the "nobody ever passes gas at a Disney park" and the Smellitzers pumping out their odors may have something to do with that. A network of tunnels allows park employees, known as "Cast Members," to scurry around and appear at just the right time and place. As explained in a delightful posting by Gabriel Oliver, "if you've ever been to a Disney park, one thing will stand out: It's clean. Thousands and thousands of people are all around you, most of them kids, and there is zero trash on the ground. No sticky gum residue, no used condoms or old panties to be seen anywhere. *Everything* is clean. Have you guessed why? Because people are popping up out of the ground to clean that shit up, like mole people, quickly disappearing back into their subterranean kingdom with you none the wiser."[296]

Never wanting to miss a revenue opportunity, Walt Disney World® offers a $79 "Keys to the Kingdom" behind-the-scenes tour that includes a visit to these tunnels, or Utilidors.

Like DARPA, the Disney Corporation has a significant research enterprise. Its motto is "The Science Behind the Magic" and they live up to it, often working in partnership with leading universities.

Disney Research has produced a Swept Frequency Capacitive Sensing technology called Touché, which combines the touch screen capabilities of a tablet with the motion sensing abilities of a device like Microsoft's Kinect. According to the Disney Research website, it "can not only detect a touch event, but simultaneously recognize complex configurations of the human hands and body during touch interaction."[297] It can detect, for example, a hand submerged in water and subtle gestures. Touché can enable smart doorknobs and replace computer devices like a mouse and keyboard with simple gestures like a finger pinch. It will undoubtedly find its way into a Disney theme park.

Another wonder from Disney Labs is "Ishin-Den-Shin" which translates as "unspoken mutual understanding." A user records a message, and can then replay it by touching the ear of a friend or classmate.[298] The stated goal of Disney Research is to make the technology invisible, because as noted by Mark Weiser, "the most profound technologies are those that disappear."[299]

There is a lot to be learned from Disney's bold, profitable foray into tagging people like farmers tag their cows. The first is how a technology can become "*de facto* mandatory." Sure, it's possible to live in Los Angeles without a car or in Beijing without a mobile phone: it's just not much fun. It's the same with MagicBands. Just wait till your kids look up and ask, "Daddy, why are we standing in line while those people don't have to?"

A second take-away lesson comes from the "long-range" nature of the wireless technology associated with MagicBands. With a good enough network and smart enough tracking software, park vendors, characters, hackers, and identity thieves will literally be able to "see you coming."

Did you buy a vegetarian lunch? Shake Minnie's hand but not Mickey's? Perhaps you visited the washroom every half hour? Sporting the Disney version of a prisoner's ankle bracelet, you're sending a constant flow of information to a system that's trying to decide what to sell you next.

Disney Chairman Thomas O. Staggs insists that all the features of the wristband can be turned off. If you don't want Cinderella to call your little princess by name, well that's your problem. Of course, you cannot opt out of the core tracking functionality of the devices used by Disney for market research.

Disney's creepy MagicBands serve as the proving ground for the "please track me" world that marketers dream about every night, as well as a laboratory for a grand experiment. Just how much privacy will you surrender for a service or status that you feel is valuable?

Of course, the minute you leave the park, you can toss that MagicBand in the trash. Or you can keep it and try to figure out how it works and how it can be tricked. I suspect several dozen people who attend DEF CON every year are working on hacking these things right now.

While Disney's implementation seems especially problematic because it is completely controlled by them and fits on wrists that are three inches in diameter, it does not take a lot of imagination to see how this concept could easily extend into the wider world.

Most people are already carrying a generic version of the MagicBand called a smartphone. Indeed, a suite of apps with names like Family GPS Tracker allow you to follow your loved ones, or anyone who gives permission, and track them on a map.

Smartphone tracking permission can sometimes be granted surreptitiously. There are many ways to put malware on phones, the latest being via "Black Widow" phone chargers that not only charge your phone, but also install rogue software to track you, read all your messages, perhaps even plan a good time to rob your house.

Indeed some educational sites like pleaserobme.com have been created to dramatize the risks of putting things online like "Whee ... we're all on the way to Hawaii. Dog's at the dog sitter's. Neighbor's picking up the mail. Back in two weeks. Aloha!!!"

It's a good thing that dog's safely at the dog sitter's, because pets that are home alone do indeed get stolen, or at least hassled by the police.

# Pet Creep

I tell this tale with a bit of trepidation because it was imparted to me by a gun-toting officer of the law in a place of liquid refreshment. However, I have every reason to believe it is 100% true and it illustrates an important point of technocreepiness: Give somebody physical access to your technology, even briefly, and they can effectively "own" you. I also know that the person who told me this story would never be allowed to put it in writing, so here goes.

My Federal law enforcement buddies needed to do a "black bag job" on a certain organized crime figure's house. This involved entering the premises to install a piece of spyware on the suspect's computer. So they sent him a nice letter from a local restaurant inviting the whole family to a complimentary meal, reservations required, of course.

On the night of this expedition, once the family was gone, my Fed friends opened the front door and were confronted with the family pets. A dog and a cat.

The dog was no problem, or at least "nothing that a nice juicy steak wouldn't take care of." They had come prepared with fresh meat. The family cat, on the other hand, dashed out the front door. Several burly Federal agents were now dispatched to hunt down the animal, staying in touch by secure radio. After a while, they reported back: "We have the suspect in custody." The keystroke recorder was safely installed, and the cat was returned. They locked the door the same way they got in. They did notice a large amount of yowling and barking as they departed, but thought nothing of it.

"The f****** feds broke into my house and switched my cat!" the man told his criminal buddies the next day. A case of mistaken feline identity, but at least the keystroke recorder did its job.

I was once hired by the CEO of a company to do a similar covert intervention on a corrupt employee's computer in the middle of the night. I discovered that the office building's lights were programmed to go off at 6 PM. I knew it would arouse suspicion if I turned them on and that fact showed up in a log. I too had come prepared, so I had a romantic candlelit session installing the evidence-gathering device. The employee resigned the following week.[300]

Today, of course, penetrations like these would be done remotely and the guy would probably have video surveillance all over his house. He would certainly be keeping tabs on his pets and possessions with a few cameras.

These days, the business of pets is booming, especially with respect to technology. There are people using technology to relate to their pets in ways that are usually reserved for kids of the non-fur-bearing variety. And, it's getting creepier by the day.

The motto of Biscuits and Baths, at 87th Street and Park Avenue in Manhattan, is "The most fun your dog can have without you." They offer an extensive program, including: "Frolic in Central Park. Eat an all-organic lunch. Take an afternoon nap. Urinate on a tree." Even at a high-end doggie day care like Biscuits and Baths, owners still fret about their pets during the day. That's why some pet facilities are sprouting the same video technology you can use to spy on your human kids in their day care.

And if you do leave your precious pet home alone, technology also offers solutions such as the Petcube. Its creator, Alex Neskin, wanted to amuse his pet Chihuahua Rocky who was left home alone. Since Rocky's favorite activity is chasing a laser pointer, Neskin taped one to a web camera that he could control remotely. He also allowed friends access to the website so that they too could play with Rocky.

There is also a simpler solution called Pintofeed. It allows you to remotely dispense food for your pets, but there's no loudspeaker to let your pet know when dinner is served. Once we accept that pets are indeed child surrogates for many people, these technological accommodations don't seem that weird. But then we move into new territory, like animal selfies.

Cats now have their own photo sharing app, Snapcat. It was reportedly cobbled together in twenty-four hours at a Berlin hackathon, and allows cats to take a self-portrait by swiping their paw across a smartphone. Clearly some people take their pets as seriously as they do children.

The real creepiness here is that we have distanced ourselves so much from our pets, both physically and psychologically, that we feel we need to make it up to them with technologies like Petcube, Pintofeed, and Snapcat. Some people are so concerned about the welfare of their pets that they even write them into their will.

Real estate tycoon Leona Helmsley left a $12 million trust fund to her female Maltese dog, Trouble, while cutting two grandchildren out of her will completely.[301] A judge reduced the pooch's purse to a mere $2 million. That probably would have covered many years in a nice suite with room service and twice-a-day walks by the hotel's concierge. However, Trouble only outlived his doting owner by three years.[302]

You don't have to be a billionaire to make thoughtful provisions for your pet. Animals can inherit wealth in most U.S. states, which have guidelines for a formal "pet trust." Other countries, like Canada, don't allow leaving money directly to dogs, cats, birds, or squirrels, but your fur kids can be provided for within regular trust structures.[303]

One Canadian "make your own will" site, www.formalwill.ca, has a downloadable form for a "Pet Guardian Agreement," providing directions for feeding, hiring groomers, walkers, and even instructions for when heroic medical treatments are to be given.[304] The pet form costs the same ($59 CAD) as the one that covers your human offspring. Perhaps the reason this is offered as an online service is that most of us would feel like an idiot discussing cat food brands in front of a lawyer who was billing us $350 an hour.

A study by scientists at the University of Veterinary Medicine in Vienna reported striking similarities between owner-dog and parent-child relationships. Researcher Lisa Horn wrote that the "unique relationship between adult dogs and their human owners bears a remarkable resemblance to an infant attachment bond: dogs are dependent on human care and their behavior seems specifically geared to engage their owners' care-giving system."[305] So dogs, at least, are genetically engineered to suck up to us to make sure they get fed, walked, and, sometimes, remembered in the will.

Our fixation with pets, often elevating them to the status of family members and surrogate children, is a source of some amusement in other countries where the concept of a yearly vet checkup or gourmet dog food is seen as risible.

An ad campaign that ran in New York subway cars in 2012 purported to offer pet makeovers, parodying the style of a well-known local plastic surgeon. The ad turned out to be a clever tongue-in-cheek promotion for Nick Kroll's sketch comedy show on Comedy Central, but there actually is a thriving business in cosmetic surgery for dogs, and not just for facelifts to correct drooping eyelids. The other end of the pooch also gets attention.

A Kansas City, Missouri, company called CTI Corporation has spared over half a million male dogs, cats, bulls, monkeys, horses, water buffalos, rats, and yes, prairie dogs, from embarrassment at the off-leash park or paddock. The company sells silicon testicular implants called Neuticles®, which are used to fill empty scrotums after an animal has been "fixed."

In his book *Going … Going … Nuts: The Story Has To Be Told*, inventor Gregg Miller says he owes the multi-million dollar idea to his beloved bloodhound Buck:

> Buck looked up at me with a puzzled expression—his Bloodhound look of worry and concern. He looked back down again—and then

> back at me a second time with an expression of "where did they go—what has happened to me?" He didn't clean himself—only had that look of bewilderment. Buck knew they were gone and for over a week seemed sluggish and depressed.[306]

Working with a vet to solve Buck's problem, Miller's work not only earned him U.S. patent #5,868,140 and a thriving business. He was also the 2005 winner of the tongue-in-cheek IG Nobel Prize in medicine, a fact he proudly trumpets on his webpage.[307]

Miller's company also does a lot of really good things, like making ear and eye prostheses for injured animals. Neuticles® Original are very reasonable: the medium size for dogs 30 to 60 pounds costs only $84 each, though you'll probably want to get the pair for $139. You might also throw in a tank top sporting the company's slogan, "It's Like Nothing Ever Changed." Though you can order Neuticles® on the Internet, they're not approved for human use, especially not the horse, bull, or water buffalo sizes.

Why do Neuticles® seem like a creepy use of technology? Surely not because, as some might suggest, it's a waste of medical resources that could have been given to humans. After all this is a procedure, paid for by the pet owner, done by a veterinarian, and taking an extra three to five minutes when they're working in there anyway. It's more about how we view the minds of our domestic animals.

The idea that our pets have fragile psyches that need nurturing does go a long way to imputing human emotions to non-human creatures. While cats purr and dogs wag their tails, and this certainly seems correlated with happiness, it is quite a leap to assume that they're having the same kind of thoughts as we do. Then again, soon we might be able to ask them what's on their minds.

On April 1, 2012, Google announced "Google Voice for Pets" which would allow you to receive text messages and even voicemails from your dog and cat. The key innovation was "Voice Communication Collars." These Google-invented devices "fit around your pet's neck and use a series of sensors to record audio directly from your dog or cat's vocal cords, using technology originally developed for NASA spacesuits."[308] They didn't stop there, going on to say "voicemails from your pet would be pretty silly if you haven't been trained how to understand cat or dog. Thankfully, we've solved that problem too. We took our voicemail transcription engine and combined it with millions of adorable pet videos from the Internet, training it to understand our furry friends. Now our transcription engine can now(sic) translate cat meows or dog growls into English!"

If you clicked on the link "to be one of the first pet owners to try our special communication collars," you were taken to a Google search on "April Fool's Day." This was one of Google's little pranks.

But, not so fast. Perhaps talking to the animals is not so far-fetched. Northern Arizona University Professor Emeritus Constantine Slobodchikoff's work on communicating with prairie dogs has been featured in *The Atlantic, National Geographic*, and *Smithsonian Magazine*.

In an interview with *The Atlantic*, Slobodchikoff moved from analyzing the "jump-yip" gesture of the prairie dog to speculating about an electronic device that might open up the hidden world inside the brain of a pet:

> I think we have the technology now to be able to develop devices that are, say, the size of a cellphone, that would allow us to talk to our dogs and cats. So the dog says "bark!" and the device analyzes it and says, "I want to eat chicken tonight." Or the cat can say "meow," and it can say, "You haven't cleaned my litterbox recently." He predicts this will be a reality "probably five to ten years out."[309]

We might be skeptical of Slobodchikoff's vision of human-animal chatter in the coming decade. But he does have an ace up his sleeve. Body language. He figures we have been wasting our time "barking up the wrong tree" in studying animal language by just trying to decode their sounds. Animals communicate with their whole bodies, including their urinary apparatus. They may use signal systems we cannot even detect. Slobodchikoff confirms what every slouching, iPod-wearing teenager is probably told before that first job interview—"when spoken language and body language conflict, the listeners pay attention to the body language."[310]

Of course, if you really want a pet to converse with, you could try a virtual pet like a Furby. The unique feature of the Furby, aside from being cuddly in an alien sort of way, was its language ability. Out of the box, all Furbys spoke their "native language," Furbish. Given a command, a Furby might say "doo-dah" (meaning "OK, I'll do that") or "Boo" meaning "no way, José." Then, according to the hype that had frantic parents tearing them off the shelves around Christmas 1998, your Furby would gradually "learn English," and shed its Furby baby talk for vaguely proper English. This worried the National Security Agency.

In 1999 the NSA issued a memo notifying its employees that "personally owned photographic, video, and audio recording equipment are prohibited items. This includes toys, such as 'Furbys,' with built-in recorders that repeat the audio with synthesized sound to mimic the original signal."[311]

The fear was that a rogue Furby could record confidential information and then blab it back if its owner stopped off in a bar on the way home with the thing in tow.

Apparently lacking the will, technical ability, or spare Furbies to cut them apart, the U.S. National Security Agency simply banned them from the premises. At a time when Edward Snowden was still trying to get his first driver's license, and Wikileaks was only a dream of its "chief visionary" Julian Assange, the NSA was fretting about stuffed toys.[312] After all, they were made in China.

We now know that first generation Furbys were faking it. They only had a two hundred word vocabulary—one hundred in Furbish, and one hundred more in English—that they were pre-programmed to reveal. In a blog post, Dwayne Hoover analyzed the inner workings of the original Furby:

> it's not going to learn or say anything different than the 100 English words it was already programmed to 'learn.' You can read it Portuguese porn articles every single day for six months straight, and it's still going to end up saying, 'I big worried.' But, it's easy to see how the NSA wouldn't know that. It's not like they are big on, you know, gathering information about things.[313]

We all fell for it, and indeed there were Christmas-time "Furby Fights" at toy stores around the world. Webpages even revealed secret "Furby cheats" like "cover his eyes three times then pat him on the back, and he will crow like a rooster"[314]

If one Furby was good, two were even better, since they did have some rudimentary ability to interact. They reportedly played Hide and Seek with each other by saying the phrase "Hey Kitty Kitty Kitty Hide" back and forth.[315] Real pets were reportedly driven insane by their incessant chatter.

This kind of annoying behavior, coupled with the lack of an on/off switch, spawned creative ways to "kill a Furby," including putting him in a microwave so "the electronic parts of your Furby would be burnt out and destroyed in a very short time."[316] Your microwave would also be destroyed, giving Furby the evil last laugh.

Hasbro brought Furby back in 2012 to bedevil a new generation. Now with LED eyes, the new Furby comes in a box emblazoned with the ominous tagline "Furby—A Mind of Its Own." This new Furby interacts with computers and smartphones and can lay eggs which hatch into Furblings.

If Furbies could be demanding and petulant, bleating out "Ah-May Koh Koh—Pet Me More!" at the most inopportune times, the other virtual pet craze, Tamagotchis, was known to interrupt business meetings in Japan and cause people to miss their turn on a golf course.

With their built-in digital meters, these pocket-sized creatures demanded regular care, feeding, and even bathroom breaks. The programming inside the Tama-Go has been reverse-engineered by Natalie Silvanovich, who posted some of her findings online.[317]

She shares Tamagotchi secrets like "from code inspection we learn that it is equally likely a girl will be Belltchi and Hositchi, and equally likely a boy will be Mattaritchi or Ahirkutchi."

These glorified digital watches, and our reaction to them, actually provides some great insights into human nature. From the hatching of the egg, the Tamagotchi's life is a series of suspenseful, if somewhat pre-programmed moments. Will it be a boy or a girl? Which kind of baby personality will it have? What factors determine the variety of Toddler or Teen it will grow into? The randomness and apparent unpredictability is evidently an appealing feature of the device.

Digging into the chip inside Tama-Go's programming, Silvanovich found that initial breed and gender are determined randomly, based on the precise clock setting at the moment the device is activated.

From that point on, Tamagotchi child-rearing imitates real life. Ignore your baby and you are going to get one of the nastier toddlers. Also, Silvanovich found, "if you care for your toddler poorly, you need to make up for this in discipline in the teen years or else you will get a 'bad' character." By pretending to be alive, yet clearly not being truly alive, the Tamagotchi falls into the "Uncanny Valley" that is both the bane and the delight of robotics researchers.

If you think the era of virtual pets has passed, you don't watch a lot of late-night TV. A fake parrot is now making the rounds with the Sham-Wows and home gym machines. Perfect Polly doesn't need food, or a cage, and never stinks up the room.

The infomercial for Perfect Polly implies that this motion-activated mechanical bird, which *does* have an on/off switch, will be the perfect companion for Grandpa since it will react to his every twitch with a tremor of its own. In the commercial, children who appear old enough to know better, accept this plastic pet as a *bona fide* and welcome addition to the family. And, thanks to an amazing feat of avian mind reading, we're even assured that a real parakeet can't tell the difference between a potential mate and this plastic imitation.

The Furby and Tamagotchi crazes of the 1990s saw people adopting robotic pets and treating them as real living things, and Perfect Polly is merely the newest iteration of this phenomenon. Our longing for companionship, even of the mechanical variety, appears to be growing ever more pervasive and complex.

# Robot Creep

If the word "robot" conjurs up Issac Asimov–style humanoids made of sheet metal, or the toy you got for your fifth birthday, it is time for some updating. The vast majority of robots today are either computer-controlled arms that weld car parts together or goofy-looking contraptions that defuse bombs or roam around Mars. Even that image is about to change.

In the near future, some robots will look like insects. Swarms of insects. Researchers at North Carolina State University are thinking of using sensor-equipped cockroaches to explore dangerous sites like collapsed buildings. The plan is to have them "signal researchers via radio waves whenever biobots got close to each other." In a press release from the University, Dr. Edgar Lobaton says, "One characteristic of biobots is that their movement can be somewhat random. We're exploiting that random movement to work in our favor."[318] They've been working with Madagascar hissing cockroaches.

Like many technologies that move from the lab to the consumer world, you can now buy the technology to make your own iPhone-controlled living being. After a successful Kickstarter campaign ($150 with "a dozen well behaved roaches"; $100 if you already have all the roaches you need), Backyard Brains of Ann Arbor, MI, offers a kit that allows you to hack the nervous system of a cockroach.[319] An ethical furor around animal cruelty has arisen, though the comment "let he who has never squashed a bug throw the first stone" seems to have calmed that somewhat.

The military has long been interested in using bugs for surveillance and as weapons. In a 2009 book called *Six-Legged Soldiers: Insects as Weapons of War*, Jeffery A. Lockwood examines the insect-created 1343 pandemic in Kaffa which helped Janiberg, the last

Mogul khan. Napoleon's 1799 and 1812 defeats were caused by plague-bearing fleas. The book also describes the use of insects as instruments of torture, even on young children.[320] The U.S. Defense Advanced Research Projects Agency has experimented with weaponized flying insects.[321]

The Japanese-originated concept of the Uncanny Valley provides clues to our reaction to robots that are just a bit too human. A great example can be found in a YouTube video called "Invertuality: Jules says Goodbye."[322] Jules is "a conversational character robot" built by scientist-artist David Hanson. Like BINA48, Jules has a human-like face made of a foam-rubber composite called Frubber. It's designed for statistically perfect androgyny. He looks from person to person, and remembers where people are so he can look at them.

In other videos, Jules expresses fear about traveling to England and curiosity about his own sexuality. In addition to having realistic facial expressions, Hanson's robots, which also include a pretty good Albert Einstein, carry on rather convincing conversations.

Artists are not the only ones making robots. At a robotics demonstration at Fort Benning in Georgia, Scott Hartley of 5D Robotics reportedly said that "ten years from now, there will probably be one soldier for every ten robots. Each soldier could have one or five robots flanking him, looking for enemies, scanning for land mines."[323]

Critics of this trend argue that the psychological threshold for going to war may be reduced if we are using disposable robots. Drones already allow us to fly missions without risking human life; robots might make soldiers virtually invincible, especially against adversaries who are not similarly equipped.

We might even get to the stage anticipated by science fiction writers where countries in conflict simply duke it out in cyberspace to see who would win, based on mathematical models, and then the proper number of citizens on each side are executed in the settling up. It would be an efficient if chilling way to handle disputes with our neighbors.

Will robots also minister to a soldier's need for companionship? Throughout the ages, where there have been (human) soldiers there have been prostitutes. Now there is serious talk about Android brothels in the future. The world's oldest profession may be ripe for some very creative automation.

New Zealand researchers have "predicted that robot sex workers will replace human prostitutes by 2050."[324] Calling it the next logical step in the billion dollar sex toy industry, a YouTube video promises lifelike sex robots with sanitary features that will eliminate sexually transmitted diseases.

In a real academic paper, Ian Yeoman and Michelle Mars of the Victoria Management School in Wellington give us a provocative, if creepy, peek into a fictitious future establishment, the Yub-Yum, located near a canal in Amsterdam:

> The Yub-Yum offers a range of sexual gods and goddesses of different ethnicities, body shapes, ages, languages and sexual features. The club is often rated highly by punters on www.punternet.com and for the fifth year in a row, in 2049 was voted the world's best massage parlour by the UN World Tourism Organisation.[325]

Would humans actually jump species to have sex with robots?

The website YouGov and The Huffington Post commissioned a survey on robosex and found that 18% of us believe that robots will have sex with humans by 2030, and 9% would go for it if they could.

In answer to the question "if it were possible for humans to have sex with robots, do you think that a person in an exclusive relationship who had sex with a robot would be cheating?" 42% said yes, 31% no, and 26% probably gave the most honest answer, "not sure."[326]

There are some deep definitional and philosophical questions at play here. What is a robot? RealTouch Interactive is a device that connects to a USB port and to a certain part of the male anatomy. It can then be controlled remotely over the Internet. Is this robot sex? Is it infidelity or just high quality porn viewing?

The RealTouch is said to bring "porn into the 4th dimension." Gizmag reviewer Loz Blain admitted that he has "been using this USB-controlled pleasure machine to have amazingly realistic long-distance sex with girls on three different continents."[327]

The female side of this market is also being addressed. In an article on Wired.com, writer Regina Lynn documented the day a UPS truck delivered her high tech Internet-enabled sex toy for women called a Sinulator. Both Sinulator and RealTouch appear to have ceased selling their products, but others are filling the void, including some that include smartphones apps.[328]

Teledildonics moved into virtual reality with the November 2013 introduction of VR Tenga, a robotic sex toy from Japan that coordinates its ministrations with the Oculus Virtual Reality headset. According to people who have tried it, the effect is more than adequate.[329]

Users of this device may be showing us the way of the future—the rise of human/computer hybrids. The VR Tenga is more than just a communication channel: it adds input of its own into the experience, but there is a human on the other end.

In Japan, "Doll No Mori" ("Forest of Dolls") charges the equivalent of about $110 for a seventy-minute "doll escort service." In a book on robot ethics, this example is discussed with the appropriate degree of gravitas. While concluding that the dolls are not fundamentally different from vibrators, the authors do raise thorny issues about whether spending an hour enjoying the pleasures of Doll No Mori constitutes marital infidelity.[330]

Considering the extreme "robots have rights too" viewpoint, this book also suggests that "natural law mitigates in favor of an artificial consciousness having intrinsic rights, and therefore, simply by virtue of having an artificial consciousness, a robot should be ascribed *legal rights.*" So much for the idea that robosex will be cheap, simple, and uncomplicated.

For decades, MIT Professor Sherry Turkle has stretched our minds about interacting with technology. In her Plenary Lecture to the 2013 AAAS conference in Boston, she described her experiences watching people interact with Kismet, a big-eyed humanoid robot that lives in a lab at MIT. Even highly intelligent people seem to enjoy conversing with Kismet.

> This is not about the robot deceiving anybody. But because of that eye contact, those facial expressions, the voice that responds to the cadences of your own, there is what I call 'a moment of more.' Talking to Kismet you have that pleasurable experience of being understood, even though you know that you are not really understood.[331]

Turkle suggests that we all crave attention, which explains the popularity of Facebook and Twitter, social outlets that provide us with "so many automatic listeners." She claims that our interaction with sociable robots will change us, causing us to rethink the meaning of words like "caring," "friend," "companionship," and "conversation." Anyone who has ever delivered a longish soliloquy to a dog or cat will probably empathize.

What limits should be placed on robotic access to our lives? Gmail got us used to the idea that robots should read our mail to sort spam and to advertise to us. If we agreed that Google can robotically read our messages to give us free email, how can we argue on principle that it's wrong for a government agency to read our traffic to keep us safe? Of course, "informed consent" is a key difference here. Yet how many Gmail users are well enough informed to really give that consent? As General Keith Alexander suggested in his speech to the 2013 Black Hat conference, many people might actually approve of the NSA's data collection techniques if they thought about alternative security measures that might be even more draconian.

Another robot that we all use on a daily basis is some sort of a search engine. Even toddlers now understand the concept behind Google or Bing or Yahoo Search. The actual workings are extraordinarily complex, involving spiders that traverse the net, indexing files and mathematical ranking algorithms. But the end results seem like magic.

The demand for this functionality is so great that it has even been implemented in places where the Internet has yet to go. MIT Media Lab co-creator Nicholas Negroponte once proudly showed me the work that his students did in rural Cambodia. Every day, someone would ride a motorcycle with a wireless access point along a road, letting it communicate with systems in schools and other buildings in order to provide once a day email and search access.

These MIT students quickly figured out that certain queries, like how to grow rice or avoid HIV/AIDS, showed up with some regularity. They "cached" those answers in the offline computers, meaning that if you asked the right question, you got an uncanny instant answer, even without connectivity.

There is little question that the existence and convenience of search engines has fundamentally changed our way of thinking and learning. In a 2009 interview with Charlie Rose, Google chairman Eric Schmidt noted that his company's signature product has made it unnecessary for today's students to do the kind of rote "education" that he was apparently subjected to.

> When I was 13, and I grew up in Virginia, I was required to memorize the 52 cities that were the capital cities of each county of the state of Virginia, which I mastered after a lot of work. Today, of course, there is a nice table in Google that tells me all that. I don't know why I'd have to memorize that.[332]

In fact, the regular use of search engines appears to actually "rewire our brains." Like any change, it undoubtedly has both positive and negative aspects. Researchers at UCLA's Semel Institute for

Neuroscience and Human Behavior put people inside fMRI brain imaging machines and found that "for computer-savvy middle-aged and older adults, searching the Internet triggers key centers in the brain that control decision-making and complex reasoning."

Compared to people reading a book, Internet searchers showed increased activity in the dorsolateral prefrontal cortex area of the brain. The researchers also managed to locate, back in the late 2000s, some people who were not experienced with Internet search and observed that, after five days of searching for one hour per day, "the subjects had already rewired their brains."[333] One of the researchers, Dr. Gary Small, has expanded on these findings in a book called *iBrain: Surviving the Technological Alteration of the Modern Mind.*[334]

The advent of visual search engines, such as TinEye and Google Image search, has also changed the way we think and search for information. Having an unlimited ability to search for anything can lead to mischief. There are actually some searches, such as for numbers in the format of credit cards, that can get you banned by Google for a period of time.

Of course, Google is not the only search engine in the world. Another, called Earthcam, specializes in helping you find webcameras that are available, and sometimes even controllable, from the Internet. Some are put there deliberately by tourist attractions like ski resorts. Others are available because they have been poorly configured or improperly secured.

Programmer John Matherly built a tool called Shodan, paying homage to a character in the *System Shock* video game series. When pointed in the right direction, it robotically traverses the Internet, looking not only for cameras but also for power plants, industrial sites with weak passwords, and just about anything in the big wide "Internet of Things" that is available to the public.

Shodan looks in places that Google doesn't go for things that people don't want you to see. One of its strongest abilities is finding systems that monitor industrial processes, and in some cases, allowing a user to control them remotely. According to a media report, Shodan users have discovered "control systems for a water park, a gas station, a hotel wine cooler and a crematorium. Cybersecurity researchers have even located command and control systems for nuclear power plants and a particle-accelerating cyclotron."[335]

Imagine an unseen hand from the other side of the world suddenly taking charge of a nuclear facility or a city's transit system. Still recovering from the impact of natural disasters like the Fukushima tragedy, some worry that a fiddling hacker pushing buttons in the wrong sequence just might take us all back to the Stone Age.

# Creep Theory

Things were both brutal and creepy in the Paleolithic era as our ancestors struggled to survive. *Homo erectus, Homo habilis,* and *Homo neanderthalensis* all had the technologies appropriate to their time: stone tools, clothing, and most especially fire. Recent plant ash and charred bone evidence from the Wonderwerk Cave in South Africa show that, even a million years ago, early hominids harnessed the power of fire on a routine basis.[336]

We can only imagine how bizarre the astounding transformation of matter by fire would have appeared to these people. They would have been as unsettled by this mystery as we are when we walk by a billboard and it displays something we just mentioned in a tweet. They figured it out, and so will we, but not without some burned fingers.

In their article on the Wonderwerk Cave discovery, anthropologist Michael Chazan and colleagues call the ability to control fire "a crucial turning point in human evolution." In a very real way, we have reached a similar juncture. Information, and the technologies that handle it, are transforming our lives in ways as fundamental as the changes brought by fire.

Since we've had information processing for over sixty years, one might think we've moved beyond the "Ugh. Look. Fire!" stage. Actually, and I can say this with confidence because I've been involved with computers since 1965, the first four or five decades of information technology, for all but the most advanced thinkers among us, were spent just rubbing the sticks together:

First we automated things that we understood, like payroll processing, airline reservation systems, and searching for stuff in the library. A few bright lights like Joseph Weizenbaum and Ray Kurzweil pushed us to think about using technology to do things differently, instead of just billions of times faster and more efficiently.

The way we applied technology in the past made eminent good sense because that's exactly what the times called for. Just as Henry Ford's assembly line made car making more efficient, the IBM 360/50 computer ensured that I got my paycheck on time and that the calculations were done right, as long as humans entered the data correctly.

Now, however, as biomedical and information technologies merge in seamless ways, we don't really know where we are going. Information will still be the spark, but our bodies and our entire lives are becoming the fuel.

It is clear that we should be thinking about the moral, ethical, and even spiritual dimensions of technology before it is too late. We know we will not get it 100% right, because some entrepreneur or hacker will always come up with something clever that we never anticipated. That's why having a framework based on past experience can help. It's time to consider why some technologies strike us as being technocreepy.

Helen Nissenbaum has written extensively about "Privacy as Contextual Integrity," suggesting that a shared understanding of the norms of information use is key to protecting privacy. As she writes, "demanding that information gathering and dissemination be appropriate to that context and obey the governing norms of distribution within it" will provide a benchmark of privacy protection.[337] Nissenbaum's ideas are explored in an article by Alexis Madrigal, which includes an excellent example of contextual privacy.

Madrigal notes that some people are offended by Google's Street View car even though they are standing in a public street and can be seen by their neighbors. "If I'm out in the street," he writes, "I can see who can see me, and know what's happening. If Google's car buzzes by, I haven't agreed to that encounter. Ergo, privacy violation."[338] The key criticism of Nissenbaum's framework, Madrigal writes, is that "it rests on the 'norms' that people expect."

To explore what contextualized privacy really means to us, here is a model that illustrates some aspects that have emerged as common threads in the examples we have considered:

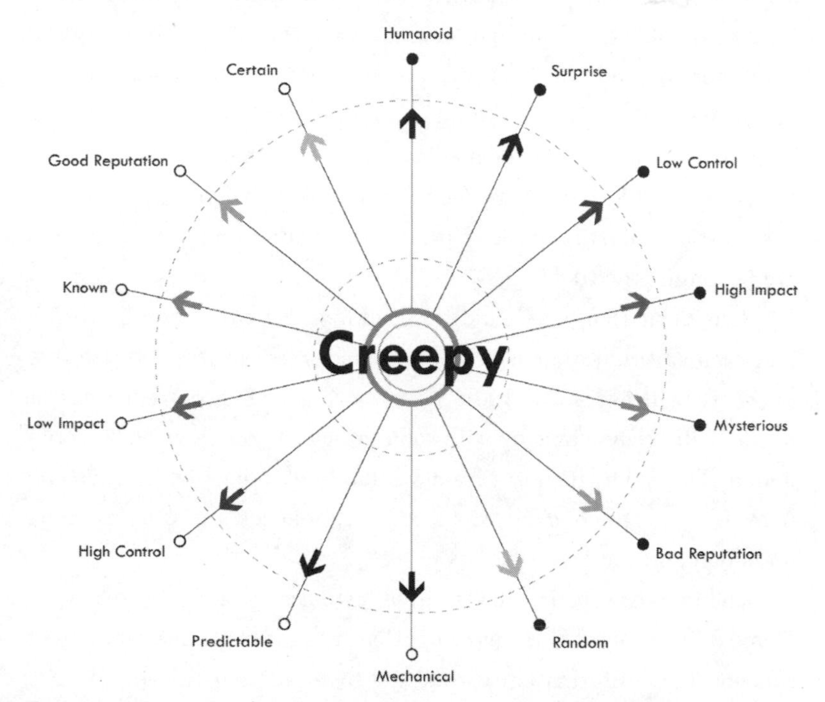


Figure 9. Dimensions of technocreepiness. Concept by Thomas P. Keenan. Image created by N.R. Dekens.

In 2012, the *New York Times* described a controversial smartphone app called "Girls Around Me" (GAM) as "Taking Creepy to a New Level."[339] While it's not quite true that GAM demonstrates every aspect of technocreepiness, it does come pretty close.

GAM allowed a smartphone user to snoop on strangers in the vicinity who had checked into the location-based Foursquare service. On the surface, that seems totally reasonable. If someone discloses their location on Foursquare, presumably they would like to be found, at least by some people. However, GAM also silently links back to the Facebook profiles of the subjects, which often contain a great deal of personal information.

So, the scenario goes, Bob, possibly encouraged by the real-time gender ratio in a bar presented on another app, such as SceneTap, sits down on a stool and orders a beer. He stealthily checks out all the women in the bar (this was gender specific, but other apps like Gays Around Me soon followed) and chooses Alice as an attractive possible companion.

Her Facebook profile, which she has not kept sufficiently private, discloses that she is not in a relationship; likes Italian cooking; and that her favorite band is The Barenaked Ladies. Her photos reveal even more details about her likes and dislikes. Armed with this conversation fodder, and with Alice totally unaware, Bob goes over for a chat …

Let's run GAM against the Dimensions of Creepiness to see how it stacks up.

1. Known vs. Mysterious. Since a person may not even know of the existence of GAM, let alone that people around them are using it while pecking at their smartphones, the odds are good that this falls into the mysterious category. Also, in 2012 at least, few people had done much thinking about how different technology platforms, in this case Facebook, Foursquare, and even Google Maps, could be melded together.
2. Random vs. Certain. It is merely the luck of the draw that Bob and Alice are in the same place tonight. Whether this technology will even affect them depends on many factors, engendering an aspect of randomness here. Humans are, by and large, creatures of habit who find random incursions into their lives *somewhat uncanny*.
3. High vs. Low Control. This is an interesting one. If Alice is a high tech wizard or a privacy expert, she may well be able to control her presence well enough to be in control of the situation. However, for most people in most bars on most nights, the

likelihood is that there is a definite imbalance of power in favor of Bob, especially since he is initiating the use of the technology.

4. High vs. Low Impact. Another situational call. Even if Bob is so aroused that he approaches Alice and says something suggestive, she might just walk away or slap him. At the other extreme, swayed by his apparent clairvoyance, she might be persuaded to go home with him. That could end well or very badly. So we come back to the important factor of intention: is someone using GAM purely as a curiosity, as a tool to build up courage, or as a means to stalk a person?
5. Human vs. Mechanical. In his landmark essay *On the Psychology of the Uncanny,* German psychiatrist Ernst Jentsch explained that "In storytelling, one of the most reliable artistic devices for producing uncanny effects easily is to leave the reader in uncertainty as to whether he has a human person or rather an automaton before him in the case of a particular character."[340]

While there was certainly human activity and input at the time of entering details on Facebook and checking in on Foursquare, that may be less true in the future. Automated data gathering along the lines of Zoominfo, combined with DeepFace-type facial recognition, may put the data in there on your behalf. Speaking at Gigaom Structure Data 2014 conference, Foursquare CEO Dennis Crowley indicated that automating check-ins is part of his company's future plans.[341] There are some Mechanical aspects at work here too. Apps can sometimes post information automatically to your social media sites. And, of course, the very process of linking up Foursquare and Facebook through a common field like email address is a mechanical activity that has some creepy aspects of human-ness.

6. Good vs. Bad Reputation. Even the name "Girls Around Me" raised hairs on the necks of many people. After being pilloried in the press, the reputation of this app went downhill fast. Foursquare

pulled the key data feed that it was using and GAM disappeared, though not before inspiring various copycats. By contrast, products with benign, even cute names like Facebook and Twitter appear to stand the test of time and work their way into our daily lives.

Some of GAM's successors may even be creepier. Jetpac City Guides sweep up photos posted on Instagram and "looks for faces in the photos, determines if they're happy or sad," writes one reviewer.[342] It also "makes style judgments (mustache? could be a hipster! lipstick? people get dressed up to visit here!)."

7. Surprise vs. Predictable. In the early days of any technology, its capabilities are often unknown and even startling to non-users. GAM undoubtedly took many people by surprise. Fortunately, it did not last long enough for people to view it is as a routine part of life in the bar scene. Other looming technologies, such as Google Glass and Oculus Rift may have enough staying power so that we will just look at Glassholes and Rifters and take them in our stride.

These Dimensions of Creepiness also provide insights into other technocreepy situations we've considered, and how the technocreepiness level can change based on various factors.

1. Known vs. Mysterious. Clearly revelations about government snooping programs fall in the mysterious category. We are told that, for reasons of national security, we are simply not allowed to know how we are being watched.

On the other hand, some technologies are only mysterious until you understand how they work. Disney's Ishin-Den-Shin communication system falls into this category. If you see people sending each other messages by touching ears with fingers, as happens in the demonstration video, it looks like magic. Once you're shown how they do it, it is still impressive but no longer mysterious.

2. Random vs. Certain. When they first started randomly arriving, those "Nigerian 419" letters were novelties and many people fell for them. When they started filling up our mailboxes, and we knew we'd get some every day, they went from being creepy to simply being annoying. Eventually we will all understand that scammers use email; that a streetlamp may start talking to us at random; and that seemingly psychic coupon we just received on our smartphone is the result of walking past a particular trash can. There is some transfer of learning: once you recognize one kind of email fraud you are more likely to spot others. But it will be a never-ending process as the bad guys get sneakier and new technologies emerge.
3. High vs. Low Control. You could, perhaps, choose not to walk down a particular street in the City of London because the rubbish bins there are monitoring smartphone pings. But why should you have to? Should hidden technology force you to alter your real world behavior? Would you even know which rubbish bin is doing this? Based on the examples studied, it seems that technology control is often more illusory than real. You can decline on principle to give your Social Security Number on a credit card application. They will simply get it from a credit bureau or data broker. The work of the Dark Patterns researchers also shows that systems sometimes make it so hard to take control that we give up. They have even coined a phrase for making a website's privacy settings ultra-complex: "Privacy Zuckering."
4. High vs. Low Impact. Potentially your health, wealth, and the most intimate details of your life are at risk here. Even little things like falsely telling a survey site that you have hemorrhoids just for the heck of it could, in theory, come back to bite you somehow in the future. We cannot possibly foresee all the impacts of the

subterranean linkages between our technological contacts, so the best policy is to treat all personal information as sensitive, and put in a liberal dose of misinformation and even deliberately misleading "facts."

5. Human vs. Mechanical. If there were (as was once the case in China) a building full of humans sorting through our Google searches to analyze them, that would be more disturbing than knowing it's done by bots. Then again, humans do have access, albeit in "anonymized" form, to the results of those bot searches and who knows what they are doing with that data. As we move into a world where human and machine intelligences merge and interact seamlessly, we will all be traveling into the Uncanny Valley. Whistleblower revelations have also shown us that data collected that is supposed to be mechanically analyzed can also come under human scrutiny. In the case of Optic Nerve, the GCHQ operation in the U.K. to intercept millions of Yahoo webcam images, it was reported that "The documents also chronicle GCHQ's sustained struggle to keep the large store of sexually explicit imagery collected by Optic Nerve away from the eyes of its staff."[343]
6. Good vs. Bad Reputation. If the "Boyfriend Tracker" folks had planned ahead, they could have named their app "Find My Phone That I Left in a Taxi" and it might not have been banned from the Google Play store. "Girls Around Me" suffered mightily from the sexist connotations of its name, whereas Facebook has a friendly feel to it. Technology creators and marketers need to do some deeper thinking about what they call their creations. Of course they shouldn't lie, but, as I noted in the MIT lab project, there's a world of difference between "Kinect of the Future" and "the Anne Frank Finder."
7. Surprise vs. Predictable. A stranger calling you by name is surprising, but not if you happen to be wearing a nametag. Especially in its early days, people armed with apps like "Girls Around Me"

had a secret weapon that gave them an unfair advantage. The same will be true of Google Glass and whatever comes next. The antidote to surprise is usually education, though, to be fair, it's rather hard to educate yourself on the impact of government programs like PRISM, Optic Nerve, and XKeyscore on your personal life, and even the inner workings of software and algorithms elude most people.

Like a lawyer preparing a witness for cross-examination, it seems appropriate to consider some counterarguments against the major premise of this book—that our lives are infected with an increasing amount of technocreepiness.

**Creepy technology can be beneficial.**

It would be negligent to fail to acknowledge that even the most troubling technologies can sometimes be beneficial to us. People have been located in remote areas because technology was tracking them, even if they didn't know it. Car accident victims have survived because an OnStar operator somewhere dispatched emergency aid even if they didn't ask for it. The Ontario woman who was reunited with her lost fifty million dollar lottery ticket probably has no problems at all with video surveillance cameras in her favorite shopping haunts.

Indeed, General Keith Alexander in his Black Hat 2013 speech tried to argue that the daily lives and activities of Americans would be much less free and more restricted if the NSA was not doing what it does, as the government would have to use other measures to counteract the terrorist threat.

The fact that we derive ample benefits from so many technologies should not obscure the real dangers behind them. With the rare exception of institutions that collect no data about us, we are almost always giving up some part of ourselves when we interact with technology. If we have learned anything from the advance of computer science in the past five decades, it is that smart people will find ways to use information, often in ways that were never anticipated. Gurus from Raymond Kurzweil to the proponents of Watson at IBM assure us that the capability of machine intelligence is about to accelerate greatly.

**Our lives are too fragmented for any system to put it all together and form a detailed and useful dossier on us.**

Some people note that their digital photos are in their camera, or burned on a CD in a desk drawer; their professional lives are conducted under one email address, and personal business under another; and if they partake in dating sites like Match.com or use Christian Mingle, they create yet another identity, probably a pseudonym. Their Amazon purchases are separated from their banking details through PayPal, and they never give their credit cards online. They infer that this gives them some degree of immunity from the creepiest aspects of technology.

That may have been an accurate picture a decade ago. Today, our photos are more likely in cloud storage or on Facebook, and our friends are busy posting photos of us. As shown by Acquisti's work, as well as studies on medical privacy by Khaled El Emam at the University of Ottawa, supposedly anonymous photos and records can sometimes be manipulated to divulge information.[344] If we learned anything from the Manning and Snowden disclosures, it is that information on us is continually being pulled together and shared without our knowledge or consent.

Some people believe that even if it is out there, extracting information about them would be like finding a specific needle in a haystack-sized pile of needles. However, this line of reasoning ignores the tremendous increase in computing power, the decrease in storage costs, and rapid improvement in data mining and analysis algorithms that have taken place over the past two decades. Another important factor is the development of keys to cross-reference us. The gold standard used to be a government-issued identity number such as a Social Security Number. Today, stores like Target are creating their own numbers for us; our email address allows us to be tied together on various sites; our browsers can be fingerprinted to uniquely identify our computers. Very significantly, the face is becoming a universal identifier and one that we cannot do a lot to change.

**We are just not that interesting, or wealthy, or scary for anyone to care about us.**

Organizations are definitely willing to make the effort to track us, and the more they do this, the easier it becomes for them. In fact, we often enter the data for them ourselves. We have also learned that everyone's Internet traffic is of interest to the security establishment and, for that matter, to some corporations.

Every time you buy a book, "Like" a Facebook post, "Friend" someone, send an email, or even log on to the Internet from a different place, you are leaving a digital trail that's being scrutinized to learn more about you. And while human relationships may come and go, your online presence is forever and can be monetized in ways you may not have considered.

**I'm not doing anything wrong, so I don't have anything to worry about.**

Just because you are innocent does not mean you are going to appear that way to authorities. A police officer once told me about a man who routinely parked his car in stall #11 of a parking lot. Unbeknownst to him, the guy who parked in stall #12 was a major organized crime figure. They often exchanged "good mornings" and that was enough to get the occupant of #11 placed in a police computer as a "known associate" of the Mafioso under investigation. Information about you can be incorrect, incomplete, or misleading. In 2006, a Canadian woman was denied entry to the U.S. because she had once attempted suicide, raising a furor that confidential Canadian medical records were being shared with U.S. agencies.[345] Investigation shows that the linkage was probably through law enforcement records.

Another woman had difficulty obtaining credit because her file stated she had appeared in small claims court. She argued that she was the plaintiff, and had won her case. However, the company holding the record refused to add that notation so she continued to suffer. These stories serve as a chilling reminder that any skeletons in your closet, no matter how ancient, may be dragged up at any time by databases that never forget.

**Computers don't really understand me so I will always be able to stay a step ahead of them.**

It doesn't matter if machines truly "understand" us; sometimes humans don't understand us either. Google Search and Gmail probably don't comprehend your whole personality in the same way as your spouse, but they know you in a different way. Your search engine may have a much better idea of precisely what you are interested in at this very moment. Since commerce ultimately happens in the present, that information is probably more useful from a business viewpoint than whether you're a Buddhist, a Baptist, a Jew, or a Jain.

Though that would not be hard to figure out either. Your metadata might show you calling or visiting a certain church, or your checking account might show a lot of $18, $36, and $180 donations, the favorite Chai numbers for Jewish donors. The other factor is that tremendous progress is being made in machine learning and artificial intelligence. As this privacy singularity meets the Vinge/Kurzweil artificial intelligence singularity ... the outcome is, almost by definition, unknowable.

In 1984, when Dr. Duncan Chappell and I were writing our CBC Radio IDEAS series *Crimes of the Future,* we picked up on some emerging crimes like identity theft and traffic in human organs that have since become household words. On the other hand, we predicted that by the early 21st century, people would be routinely using electricity to stimulate the pleasure centers of the brain for recreation. Many new drugs have emerged since 1984, but "wireheading" is still a fringe technology. And, of course, in writing the programs, we completely missed the importance of a little something called "The Internet," which was then in its infancy.

In the inevitable gap between my writing these words and you reading them, new and even more disturbing facts about technology will have emerged. We will know more about what the NSA can do. We will learn more about how companies are burrowing into our psyches in the interest of competitive advantage. More smartphone stalking applications will appear. The whole area of biological manipulation will probably grow at an exponential pace.

Some of tomorrow's headlines will be more extreme examples of creepy technologies described in this book. Others may take us in entirely new directions. Only in retrospect can we know which ones are so creepy that people simply refuse to use them. And, of course, some will be well beyond our control as individuals, and require a higher level of intervention. Many of the creepiest aspects of new technologies will be hidden from our view, and we will only catch the occasional glimpse of them.

Perhaps the creepiest aspect of our relationship with technology is the misguided belief that we can have the benefits of new technologies without the risks. Just as there is no pleasure without pain, and no peace without war, we will always need to question the cost, the risk, and the motivation of those who may benefit from changing our lives through technology.

It will not be easy. One thing is certain: we will need to continuously make decisions on both the individual and societal levels. Long before there were viral cat videos, there was a newspaper ad offering "Free to Good Home" your choice of a playful kitten or a "handsome husband, good job, but says he doesn't like cats and the cat goes or he goes." The ad then suggested "come and see both and decide which you'd like." It is almost certainly a good natured joke between spouses who really care for each other. Yet taken literally, it nicely sums up the intensity of the hard choices we will need to make:

- Every time we post a photo on Facebook, retweet on Twitter, buy a plane ticket online, or even search for one, we are making personal privacy choices that have consequences.
- When we vote for political candidates with particular views on the issues raised in *Technocreep*, we are making societal decisions that have consequences. We can also vote with our wallets, choosing the most privacy-friendly technologies.
- As we talk to our friends, our co-workers, and our children about technocreepiness, we are taking a stand about the kind of future we want to see.

If you are left with feeling that the world is spinning out of control, you've been paying attention. There is, however, some very good news. There are some concrete ways to minimize the impact of invasive technologies on your life and those of the ones you love.

But you have to start now. Let's do exactly that.

# Anti-Creep

Picture yourself suddenly living in the most repressive, repulsive, privacy-unfriendly country on earth. Imagine how such a society could nefariously use the technologies that we so readily embrace against us. How those in charge could track us, hunt us down, link us to our friends, draw subtle inferences, and, ultimately, make our lives virtually unlivable.

Think how much you would regret posting certain photos, connecting with people online, and making certain purchases. How incriminating your travel history might appear. Perhaps you once did a Google search on a politically incorrect topic or even looked up "how to kidnap a celebrity" while watching a crime show on TV. You're in big trouble now.

Most people still behave as if we live in the polar opposite of this world—a free, open, democratic society where we can placidly enjoy the many advantages of new technologies. The truth is that we are actually somewhere between these two paradigms, and slipping rapidly towards the scarier one. There are things you can do to throw people off your cyberscent and retain as much privacy and freedom as possible, but you do need to start taking action now.

In researching and writing *Technocreep*, three things became very clear:

1. **The technological capability to spy on us is becoming more sophisticated, less expensive, and more widely available.** Any shopkeeper can buy a $50 camera to watch and record the actions of customers. Malls and even whole cities can grab information on us from the unique identifiers inside our cell

phones. Facial recognition technology is progressing so fast that soon you won't need to give your name: your face will identify you at the cashier's till or even in a crowd.

2. **The motivation to use this technology is becoming more and more compelling.** Governments, corporations, and even nosy neighbors are starting to feel that "more is better" when it comes to data, and "if we can collect it, we should." The cutthroat competition between online, brick-and-mortar, and hybrid retailers has fueled an "anything goes" mentality of massive data collection, archiving, resale, and brokering. A new breed of "data scientists" is starting to rule the roost, as businesses smell profit in techniques like collaborative filtering, long tail personalization, and sophisticated suggestion algorithms.
3. **There is a tremendous variation in how well individuals guard their privacy.** Some people are using email anonymizers, Tor (The Onion Router), and multiple layers of post office boxes to cover their tracks. One fellow I know was prevented from accepting a volunteer position at a charity casino because he refused to divulge his (real) birth date for a background security check. We are now learning that these commonsense privacy protection techniques may be of dubious utility because governments, data brokers, and others have already amassed huge amounts of information about us. At the other end of the spectrum, some people are sitting ducks for identity theft, posting tagged facial photos and giving up their names, addresses, email, and other personal data in return for paltry rewards like a trade magazine subscription or a store loyalty card.

I hope that *Technocreep* has alerted you to the profound and fast-moving technological developments that are affecting your life today and will shape it in the future. This chapter is intended to give you a basic toolkit to carry out your personal action plan to deal with technocreepiness.

Please think of this chapter as the starter toolbox that you would give someone moving into a first home. Those pliers and hammers and wrenches may help hang a picture, and even go a long way towards preventing serious household damage from a leaky toilet.

However, a set of hand tools probably will not solve major maintenance problems, just as these tips will not make you immune to serious efforts to attack your privacy by governments and corporations. But you will be much better off than your neighbor who has nothing.

One difference between home repair and protecting your digital life is the rate at which new tools are required. The only hand tool I've bought in the last five years is a Torx screwdriver to deal with those tamper-resistant screws that keep popping up on things I'd like to open. If you let your digital toolkit sit idle for five years, it would be laughably out of date, as new threats emerge, and new countermeasures are developed. This information in this chapter will therefore be updated as appropriate at www.technocreep.com.

Let's get started!

**Know Thine Enemy.**

The first step in formulating an intelligent response to growing technocreepiness is to understand who is after your information and why they want it. As discussed earlier in this book, the answer turns out to be a dizzying array of governments, corporations, even non-profit organizations and snoopy individuals. They get your information directly from you, from your activities in the online and physical worlds, from each other, and from data brokers.

The revelations of Edward Snowden, William Binney, and the journalists who reported on them showed how governments are actively collecting information about billons of people and actually using it on a day-to-day basis.[346] We have even learned that the National Security Agency is using radio waves to spy on computers that are not even connected to the Internet.[347]

The examples in *Technocreep* have, I hope, convinced you that private companies can be just as curious about you as government agencies, and just as effective in their snooping, largely because they have enlisted you as their unwitting ally.

The Internet has also put tremendous power into the hands of private citizens. Your nosy neighbor might learn a great deal about you by researching you online, flying a tech-laden drone near your home, or even just gazing at you with Google Glass.

Because the stakes are so high, we are locked in a never-ending battle between increasingly sophisticated tracking techniques and clever countermeasures. Companies started to leave "cookies" on your computer to track your repeat visits. Consumers got smart and installed "cookie crunching" software. Companies then discovered "browser fingerprinting," which takes advantage of the near-uniqueness of your computer, based on esoteric things like its clock speed and what software you have installed.

Here are some things you can do to find out who is watching you in the online world and how they are doing it:

*Install software to help you track the trackers.*
Ghostery works on most browsers and mobile devices. It will show a list of trackers on most webpages and allow you to selectively block them. The Electronic Frontier Foundation has an excellent, and safe, add-on for the Firebox browser called Lightbeam that graphically visualizes the web of connections triggered by something as simple as visiting the website of the *New York Times*.

*Block pop-ups, ads, and invisible websites.*
Software to do this is widely and freely available. The most popular ones are currently Adblock Plus and Disconnect. As a delightful side effect, blocking online ads often speeds up your Internet browsing while decluttering your screen. If you are an advanced user, and using the Chrome browser, have a look at the HTTP Switchboard plug-in, which gives you even finer control over who can do what to your computer.

*Go deep diving into places like your Facebook profile.*

Often, information that has been collected about you is stored on a far-away server, so you have only indirect access. Facebook certainly operates that way, refusing to provide tools to answer simple questions like "who's stalking my profile" or "who has unfriended me." By contrast, LinkedIn and Academia.edu tell you who has viewed your profile, though sometimes you need a paid account to get the full details.

An example of indirect snooping would be to view the Page Source of your Facebook Profile. Among other things, it will have a bunch of nine digit or so numbers that correspond to people. A WikiHow article claims that you can use these to find out who is viewing your page.[348] My own experimentation, and a few off-the-record chats with folks at Facebook, suggests that things have changed since this method worked. After all, trying to torture your Facebook profile to give up hidden information is not a supported feature of the service.

The other thing people often want to track on Facebook is being "unfriended" by someone. That's a lot easier. You can simply look at your friends list and see who has gone missing. This means they either unfriended you or de-activated their Facebook account. Since this can be tedious, there are apps like like "Unfriend Finder" to automate the process.[349]

*Periodically clear out your computer's history, cache files, and other digital detritus.*

Web browsers have a "clear history" button and the best choice is probably "clear everything." This might mean you have to re-enter something like a bank account number later, but would you rather have it stolen by a hacker? In fact, if you can live without having your browser remind you where you surfed earlier today, simply tell it to "keep no history." Programs to scrub down the rest of your computer are available, both for free (e.g. CCleaner) and for a fee (e.g. Privacy Eraser Pro), and you should use one or more on a regular basis.

**Organize Your Digital Life Compulsively.**

Knowledge definitely is power in our information-rich environment. Having a transaction receipt, copy of an email, or log of a phone call can make all the difference when you are dealing with a company or institution that already knows a lot about you. Excellent digital record-keeping is also invaluable if you have the misfortune to become the victim of identity theft.

Fortunately, computer storage is trending towards zero cost, so there's no reason you can't archive your data, just like major corporations do. There is also a wide range of software, such as NeatReceipts, Mint, and Evernote that will assist you in the process of organizing your digital life.

Yes, there are privacy risks associated with each of these. Mint asks for the passwords to your banking accounts. Only you can decide if you trust Mint (owned by Intuit, the dominant makers of tax software) enough to provide that information in return for the benefits offered. In a similar fashion, you might use NeatReceipts to scan all sorts of sensitive data onto your laptop, then leave the whole thing in a taxicab or at airport security.

Here are some tips to consider:

*Password protect all your devices.*

It just takes a few extra seconds and will deter most casual intruders. It also can have some legal force because, depending on where you live, it can be a criminal offense to try to break into your locked device. Also, you should set your phone to lock itself after a reasonable period of inactivity.

*The next level of protection is to encrypt your entire hard disk.*

This is especially relevant to laptops that are traveling with you, since it will make it difficult if not impossible for someone to access the data on it. The dominant current operating systems, Windows 8 and OS X for the Mac, can do this automatically and seamlessly for you. If you are running something else, there are a range of free (e.g. TrueCrypt) and commercial (e.g. Checkpoint Full Disk Encryption) products to consider.

*Choose your password well.*
Experts advise using complex passwords, and certainly not words in the dictionary of any language. Those can be compromised by the automated brute force attack of just trying all the words. You should also use a different password at each site since there have been high profile password compromises of companies like Adobe and LinkedIn. These can put you at risk if you have used the same password on other sites. If you have problems remembering passwords, consider either a hardware or software password manager. These have their own risks but, when used properly, can certainly help protect you.

*Never put passwords, credit card numbers, or anything else that is truly sensitive into an email.*
Aside from the ability of governments and corporations to snoop, emails are favorite places for thieves to try to mine your data. They can go through, finding passwords to other accounts you have set up, and even changing them so you don't have access anymore. The day you rush out of the airport lounge without logging off from your email account might be the day that a lot of your sensitive information is stolen and possibly sold online.

*Beef up your authentication.*
Businesses have long used "multi-factor" authentication such as key fob security tokens that display a changing number that must be entered along with a password. This builds on the idea that security can be enhanced by a combination of "something you know" (e.g. a password), "something you have" (e.g. a key fob), and something you are (e.g. a fingerprint or hand geometry).

You can use your smartphone as a kind of key fob, at least to protect your Google accounts. Just turn on that company's optional "two-step verification." Then, when Google's computers see you coming from a new device, they'll require an access code that's sent to you via your phone.

*Get ready for biometrics.*

The 2013 introduction of the iPhone 5s with TouchID fingerprint was expected to bring biometric identification squarely into the consumer mainstream. The fact that it was hacked within days by using a photograph and a fake finger demonstrated that the problem of secure and convenient identity authentication will be with us for a long time. However, new products coming down the pike, such as the Myris eye scanner, will soon bring biometrics to the masses.[350]

The Sandy Hook school shootings inspired the creation of the Smart Tech Challenges Foundation, whose goal is to reduce gun violence through technology innovation. It is already spawning practical projects such as a gun that reads the owner's fingerprint and one that can only be fired by the person wearing an RFID-chipped wristwatch.[351]

*Get a digital shredder.*

Actually there are plenty of programs that have this capability, even for free, so it's no big deal to wipe files, disks, etc. The getting organized part is a bit harder. There are rules (often a seven-year retention period) for financial records, but what about your old love (e-)letters? Your baby's first scrawlings on a tablet? All those family photos that you really never liked?

For non-financial data, this is a totally subjective decision based on weighing the pros (someday you might want to write your autobiography or look back fondly at that lost love) vs. the cons (you're running for office and somebody finds a politically incorrect rant in your old files). The main thing is to think about data retention in an organized way, and of course, wipe clean (or physically destroy) any digital media before it leaves your home or office for the trash or the recycling depot.

*Build yourself a sandbox.*

For years, software developers have isolated test and production systems so that their inevitable mistakes won't bring something like an airline reservation crashing down around them. You can find software (e.g. VMware and Sandboxie) to create an isolated environment on your own computer. Another approach is to just wipe everything off a soon-to-be-retired machine and re-purpose it as your "sketchy machine." Use it for anything you think might cause security problems, after having loaded it up with virus checkers and other anti-malware software. Don't use it for anything important or sensitive, and be ready to wipe and "re-image" it. The downside of this approach is that you might have an infected machine on your network for a period of time.

**Guard Your Digital *Persona* Like a Hawk and Cover Your Digital Tracks.**

In the 1950s, some parents would send their children down to the corner store for a jug of milk and a loaf of bread with instructions to "put it on our account." Don't try that today! Instead of your family's reputation, your identity and ability to function in society is now tied to an impersonal, automated, and, it appears, quite vulnerable system of numbers and codes.

No matter where you shop, from Target to Neiman Marcus, you run the risk of hackers getting access to your credit or debit card data and other information. Both of those retailers, and countless others, have been the victims of hacker penetrations. That type of activity is beyond your control, but there are some commonsense tips you can use:

*Prefer credit over debit cards.*

You are definitely safer using a credit card than a debit card because, at the time of purchase, you are spending the card issuer's money. They are keenly interested in protecting that, and have elaborate anti-fraud measures. In most cases, you'll have zero liability for unauthorized credit card transactions. If your debit card is hacked, you run the risk of your bank account being emptied, and a protracted fight with your bank to prove it wasn't you who did it.

*You could pay cash (or Bitcoin).*

Sure, there's a risk that you'll be robbed in the street, and you will miss out on credit card perks like frequent flyer miles and extended warranties. But immunity from hacking and protection from credit card fraud may outweigh these benefits. Then again, cash may be on its way out. Just try to use it to buy a drink on an airplane, or even to pay your telephone bill.[352] Even Canada's Passport Office now refuses to accept cash. As for Bitcoins, and other digital currencies, great idea—but good luck checking into a hotel or renting a car with them.

*Monitor your accounts online regularly.*

Almost all financial institutions provide the option of online access to your account, and that has major advantages. Printed statements in your mailbox or your unshredded trash can be a gold mine for identity thieves. In fact, if you are not 100% confident about the security of your postal mail, you might consider having the physical credit cards shipped to you care of your bank branch. That little extra effort could pay off in increased security.

Assuming you have set up online access, it's possible to check your accounts regularly, and you definitely should. The earlier you catch something amiss and report it to the financial institution, the safer you are going to be. During the 2013 scandal over hacker penetration of retail giant Target, one CNN security expert urged everyone who had shopped there to cancel each of their credit cards and request a new one because "you'll have it in two or three days." Of course, if forty million or so customers actually took his advice, it would be more like two or three months. Still, if you have any suspicion that your card is being misused, it is better to be safe than sorry.

*Set up a Google Alert on your name.*

You can use the power of Google to keep a close watch on what is being said about you online. Just go to google.com/alerts and put in your name (and any variants) in quotation marks. Sure, you'll get some false hits. I know far too much about a musician and a golfer who share the name "Tom Keenan." Still, if someone is ranting about you online, this might bring it to your attention.

*Use a privacy-friendly search engine.*

The business model of major search engines, and certainly Google, is to serve up advertising to you, preferably for things you might want to buy. Advertisers pay for the privilege, and your eyeballs are even more valuable if you have been profiled and can be targeted.

Simply typing terms into the box of a search engine can have serious consequences. An accused killer's admission that he did a Google search for "how to dissolve a body" certainly didn't win him any friends in a courtroom.[353]

Perhaps the best known search engine that promises not to track you is duckduckgo.com. The folks behind it give a fascinating illustration of how search engines leak personal information about you based on your searches on www.donttrack.us. They illustrate how a Google search for "herpes" is sent to Google along with your browser and location, which may be used to identify you.[354] This can influence the kind of ads you are shown, as this information follows you around online. Just as you start that big business presentation—ads for herpes treatments appear.

Creepy tracking by search engines is not just a theoretical vulnerability. Real people have complained about having their privacy invaded in creepy and disturbing ways.

A man in Canada searched on Google for "CPAP" (continuous positive pressure airway machines, which are used for sleep disorders). In a later surfing session, he was looking at a comic strip that had nothing to do with the medical device and was creeped out to see ads for CPAP devices displayed by Google.

He filed a complaint with Canada's Privacy Commissioner who ruled that "Google's online advertising service used sensitive information about individuals' online activities to target them with health-related advertisements, contrary to Canadian privacy law." Google promised to mend its ways.[355]

*Check your environment for things that should not be there.*
Earlier in *Technocreep*, we learned about CreepyDOL, an unobtrusive $57 snooping device that someone could plug into the wall at your favorite coffee shop, airport terminal, or public library. The odds are good it would sit unnoticed there for weeks, intercepting everyone's Wi-Fi traffic. While you may not have the technical expertise to sweep for bugs, there's nothing wrong with asking "what's that?"—whether it's a box plugged into an outlet or some new icon on your smartphone.

**Be Info-Stingy.**
Many stores routinely ask for your postal code, telephone number, or some other piece of identifying information at the checkout. Savvy Canadians give out H0H 0H0, a valid postal code that happens to belong to Mr. and Mrs. Claus. Americans, of course, tend to rattle off 90210 as their fake zip code. A California woman successfully sued retailer Williams-Sonoma, Inc. for demanding her zip code, then using it to locate her home address.[356] The main reason to "just say no" to that checkout clerk is to continuously remind yourself to be very stingy in giving out any personal details.

Your refusal here is, however, largely symbolic. Yes, you are throwing a small wrench into the store's data gathering system.

However, if they want more information about you, they can go to data brokers who are happy to sell your details. Or, as Target clearly did in the case described earlier in *Technocreep*, the retailer can simply build its own profile of you, adding data to it every time you use a debit or credit card, subscribe to a mailing list, or redeem an offer of some sort. In the future, if society allows it, stores might use facial recognition or even a TouchDNA test to figure out your identity and track you.

Here are some ways to be properly parsimonious with your information:

*Give out any phone number but your own.*
Actually, it's probably best to have a small list of bogus numbers memorized so you don't find yourself at a store trying to return something and struggling to remember what phone number you gave when you made the purchase.

How to pick your fun number? You might want to think like a movie scriptwriter, and give out a number with the 555 prefix. Since the 1960s, the film and TV industries have been encouraged to avoid inadvertently showing a real subscriber's number. The numeric range 555-0100 to 555-0199 is officially reserved for fictitious numbers in most North American area codes.

*Or you could be a little cheeky in your choice of fake phone number.*
For a while I passed out the direct private line of a government minister who was in charge of protecting consumer privacy but who didn't seem very interested in doing that. The adult approach, of course, is to ask "why do you need that?" but who wants to argue with a cashier when you are in a hurry?

*But wait, I just might want to receive a phone call from those people.*
It's hard to imagine why you would actually welcome a marketing call, but if you do feel that way, there is a simple procedure that still

protects your privacy. First, create a brand new Google Account. Then (currently this will only work for users in the U.S.) create a free Google Voice number linked to it. You can then check it periodically for voicemail, or if you really want the calls, forward it to a real telephone number. You will still have the option to undo this at any point in the future, sparing you an eternity of pesky calls from telemarketers.

*Telemarketers.*

Many countries have "do not call lists" that often fail to work properly and are even used by spammers in faraway places as lists of people to call.

Having little faith in Do Not Call Lists, I created a script for having a little harmless fun with telemarketers. It was vaguely inspired by the legendary "Angel of Death prank call" in which the person called tells a cemetery plot telemarketer that he's been "thinking of taking my life, and your call is the sign I've been praying for." In his version comedian Tom Mabe even asks the hapless marketer if they offer financing for the plot, though of course he plans to need it right away.[357]

After ascertaining that a caller wants to paint my house, remotely diagnose (and hack) my computer, or clean my furnace, I express great interest but remind him or her that "You have called a premium number." Often they will persist with their script so I repeat this until I have their attention. I patiently explain that "we charge for incoming telephone calls. It's $75 for the first ten minutes and we take Visa, MasterCard, and American Express." Once I did have some poor woman offer me her MasterCard number but I wouldn't accept it. Usually they hang up, probably flagging the number in their database as "crazy person" or something like that.

If a marketer asks for you by name, don't say you are deceased, tempting as that might sometimes be. One woman did that and her credit card company canceled her card. Do not be abusive to the telemarketers; they are only doing their job. Also, it has been reported that annoyed call center agents sometimes retaliate for rudeness by passing your number around the room for everyone to call.[358]

There are also various hardware devices that claim to cut down on unwanted inbound phone calls. Perhaps the most famous, the TeleZapper, plays an "intercept tone" to tell the bad guys that your phone is disconnected. Of course, this can have its own unintended consequences. One customer who reviewed the device on Amazon .com said that his credit card company called to say his payment had not gone through, but got the "disconnected tone." This "landed me in some hot water" he reported. The TeleZapper had other quirks, like playing an annoying tone on every call, and anyway telemarketers soon figured out ways to defeat it. Likewise for devices that require friends and family to have a PIN code to make your phone ring. They sound good in principle but are pesky in reality.

*I'd like to be the President of the United States.*

Well, I can't help you with that, except to mention that the official phone number of the White House is (202) 456-1111 and it's trivially easy to make a call that looks like it is coming from there. Sites like www.spoofcard.com disguise your caller ID and even allow you to change your voice. The main lesson is that someone can do this to you. So just because the caller ID shows the name of your bank, it is not necessarily your banker on the phone.

*How old would you like to be?.*

Some teenagers are masters at constructing bogus birth dates for everything from joining websites with a minimum age to buying cigarettes and lottery tickets. While you may not want to adjust your age by decades in either direction, who is to say you can't move it a few months in the interest of privacy?

Here's an interesting experiment. Look up someone you know personally who is notable enough to be on Wikipedia. Chances are good you'll find their full birth date listed. However, if that person is a computer security or privacy expert, like several I checked, the date is quite likely to be a fake one. One expert even asked me not to mention his name in conjunction with this point because "it will just provoke somebody to try to find the real date and change my entry."

Is it OK to fib to Wikipedia? The site's "biography of living persons" privacy policy states that they will show the exact birth date if it has been "widely published by reliable sources" and notes that if the person objects then just the year should be used. Even if you're not Wikipedia-worthy yet, it's a wise move to have a bogus birth date handy for non-official purposes. Your government and bank will still demand the real one, but for most other uses any reasonable date will do. It's amazing how many online "happy birthdays" I get at the wrong time of each year because of this policy. The same goes for your address, mother's maiden name, and email address.

*Keep your body pure.*

There's emerging evidence that tattoos are a cancer risk, and not just because they can mask moles and other skin lesions. New research shows that tattoo inks, which are largely unregulated, can contain nanoparticles which may accumulate in the kidneys and other organs.[359] So, as explained earlier in *Technocreep,* if your boss comes after you with a tattoo gun to apply your new password tattoo, you might want to head for the door.

However, the biggest risk of "body art" may be to your privacy. Databases of arrest records routinely describe "identifying marks" and it is best to have as few as possible. As far back as 1959, according to a news report, there was a file with over "200,000 people arrested each year by the Los Angeles Police Department, 90,000 of which are tattooed. Each person is indexed with identifying information including a description of his or her tattoos and location on their body."[360] In an interesting twist, according to the LAPD's current webpage, sporting visible tattoos can disqualify you from becoming a member of that force.

Fast forward to 2012, when the FBI announced plans to add "scars, marks and tattoos" to its Integrated Automated Fingerprint Identification System (IAFIS), which they proudly describe as "the largest biometric database in the world."[361] Law enforcement agencies won't just be matching and tracking tattoos, they're going to try to understand their meaning "to help establish whether an individual is associated with a particular gang, terrorist organization, or extremist group."

*Try to control postings of your face (and other distinctive features).*
Of course your face is your most identifiable feature and it is pretty hard to completely control where photos of it are posted. Friends can tag you on Facebook; police can take booking photos that wind up on mugshot sites. Earlier in *Technocreep* I described countermeasures like tagging inanimate objects with your name to throw people off your digital scent. You might also tag lots of random people as yourself, making it hard for someone to guess which is the real you. Of course, this is moot if you have a perfect headshot of yourself as your profile photo and lax privacy settings on Facebook.

The reality is that, unless you are willing to eschew all forms of social media communication, keeping your face private is going to be an uphill and ultimately futile battle. The best you can do is avoid posting photos that might come back to haunt you. That session of doing tequila shots from lab glassware in high school chem lab might cost you a lab assistant's job later on in college. Those racy office party pictures probably belong on a USB stick in your desk drawer. There are countless websites where you can watch other people behaving inappropriately—there's no need to add to the supply.

Posting photos of your body, and heaven forbid, sex tapes, is another no-no. Let's just say that image recognition technology is not limited to your face—other parts of your body can also compromise your identity. A number of court cases, including the famous one involving Michael Jackson, have hinged on non-facial identification.[362]

*Tell your devices to be less promiscuous.*
No matter how diligent you are in protecting yourself, your laptop computer, smartphone, and other tech toys may subvert your privacy efforts by automatically connecting to rogue hotspots, nearby Bluetooth connections, and NFC (near field communication) devices.[363] Sure, they're trying to be helpful by reaching out, but it's far wiser to turn off the automated connections and make your devices "non-discoverable." Then just connect manually when you really want to share something. Turn off location services except when you really need them, and periodically wipe out all those airport and hotel networks that you no longer need to access.

**Create Another You ... Or Many!**
Earlier, I described how some shoppers have posted the barcode from their Safeway loyalty card online so that people could use it to obtain the benefits like price discounts without giving up any of their own personal information.

One Internet rebel who did this, Rob Cockerham, added to his already considerable fame on the net and even managed to profit modestly from his prank. After posting his barcode online, and encouraging others to copy and use it, he sold his now-famous physical "Mint Safeway Card" on eBay for $21.53.[364]

While a Safeway representative shrugged off the impact of this "odd sort of prank" in a media interview, if enough people did things like this, it might actually have an impact on their big data analytics.[365]

There are definitely times when having "another you" can be an important tool to protect your privacy:

*Make up a "straw man" for medical tests.*

New testing techniques are revolutionizing medicine, and also raising major privacy issues. Your test results may reveal current diseases, genetic predilections to future diseases, and even facts about your close relatives. Of course, if you are in the hospital and they send your blood down to the lab, you don't have much choice about how it's identified. But increasingly, tests are coming out that are optional and driven by the consumer's choice and even curiosity, not medical necessity.

One U.S. company offers (subject to restrictions in a few states) a "Comprehensive Wellness Profile" of over fifty tests that they claim provides "a thorough Biochemical assessment of your health, and includes the basic cardiovascular tests as well as diabetes testing."[366] There's also a whole menu of à la carte tests that give a whole new meaning to "playing doctor." Think you've been exposed to zinc? They've got a test for that, priced at $37. By default, the results are only available to you, in online form. If you want them sent to a health care provider, you have to specifically request that service.

Direct-to-consumer genetic testing is booming and, even with some bumps in the road, is certainly around to stay. The laws about how companies and governments can use your medical results are still being written and debated, and, as we saw earlier in *Technocreep*, they vary widely around the world. But one thing is certain: they cannot discriminate against you if they don't know who you are. So, when submitting voluntary medical samples or other data, you may want to consider becoming someone else. Such as your pet dog, cat, or chinchilla.

*But is it legal to use a false identity?*[367].

The laws about this vary depending on where you live. In general, it is illegal to impersonate a real person but not to create a fictitious identity for yourself. A further caveat is that you may run afoul of the law if you use your false *persona* to defraud someone or if your false identity is that of a public official such as a police officer.

The New York Penal Law, for example, makes it a crime to impersonate someone else and then do "an act in such assumed character with intent to obtain a benefit or to injure or defraud another."[368] So, regarding those fake name medical tests, unless you are submitting the bills to an insurance company, or lying on a job application, it's hard to imagine that you are gaining any benefit other than protecting your privacy.

*Yes my name is I.P. Freely and this is my prepaid credit card.*

Prepaid credit cards were introduced to serve the "unbanked" as well as provide "financial training wheels" for children and teens too young for the real thing. They also play a function as gifts when you can't think of anything better, and are even doled out as incentives for attending boring online webinars. You can buy them in grocery stores, drug stores, even gas stations.

The best thing about prepaid credit cards is that they are not linked to your real identity. You can pretty well be anyone you want to be, living anywhere. One particularly popular use is for Canadians to impersonate Americans to buy things, e.g. in the U.S. iTunes store, which is restricted to U.S. residents by IP-address geo-fencing and other measures. The founder of the website howtogetitincanada.com explains this and other hacks, and notes that "all of our guides were written using Canada only as an example. The same steps should work from anywhere outside of the USA."[369]

The list of things that people would prefer not to have on their personal or business credit card bills is as long as the human imagination. Porn sites. Skimpy lingerie for a mistress. Locksmith equipment. Industrial chemicals. Even if none of those appeals to you, perhaps you enjoy the occasional trip to a fast food outlet. I have heard of one executive who pays for his Quarter Pounders and fries in cash. The day may come, he reasons, when people will lump eating fast food, or even consuming meat, in with smoking or heavy drinking—still legal but socially undesirable. He is savvy enough to know that his digital trail might someday betray him.

*Have an email for every purpose.*

Nothing is worse than missing an important work-related email because it was buried in a bunch of marketing spam. If you have this problem, you probably don't have enough email addresses. While most email services provide ways to tag or categorize emails, having multiple accounts is an even better way to ensure separation.

That way you will not get cross-over between emails going to your child's PTA and your R-rated family photos. Don't laugh; a school official was suspended because a link in his wife's email signature file contained a link to a private (but not private enough) family photo album with tasteful, non-frontal nudity. For the same reason, you are advised to create a brand new email if you plan to look for love on sites like Match, Tinder, AshleyMadison or even ChristianMingle.

It is worth noting that when you create accounts, for example at Gmail or Outlook.com, information about your IP address and other data is recorded by the email provider. The same thing happens when you sign up for services such as investment discussion bulletin boards. I have been an expert witness in a court case where one such company was able to produce detailed logs of which IP address posted which comment, and exactly when it was posted.

This should not be a problem unless you do something to attract the attention of law enforcement or somehow get swept up in a government surveillance program. Since the purpose of this guide is to help protect your privacy, not to assist with illegal or nefarious activities, with any luck you will never be in this position. However, by simply doing a Google search on "anonymous email" you can find many providers that will give you a disposable email address if you do ever need one of those.

*Could I make my doppelgänger's online profile more credible?.*

Assuming that you wanted to do this, there are plenty of ways to build credibility for the fake person you're creating. You could very likely turn that prepaid credit card you've created for I.P. Freely or Seymour Butts into a legitimate-looking Zoominfo profile. Facebook probably wouldn't object either—after all you're not trying to create a profile for a cat or a fetus.

But why stop there? Your fake friend could become famous. A Los Angeles-based hoaxer named Ken Tarr claims to have talked his way onto eight different reality shows in five different cities. In a delightful *Village Voice* article about him, he explains that he calls producers with semi-believable stories and, if they reject him, he simply calls back with another crazy yarn. He advertises on Craigslist for confederates to join him on the show, and they split any winnings.

So, as writer Graham Rayman explained, "for Judge Joe Brown, he pretended to be a drunken Gypsy clown who trashed a bathroom at a kid's birthday party ... for *Unfaithful*, a show produced by Oprah Winfrey's network, he was an international security expert who was cheating on his girlfriend—who was also cheating on him."[370]

*You mean people lie on the Internet?*.

Online disinformation is rampant, with holy wars being played out daily on Wikipedia and other sites. One person puts in a "fact," which is then challenged and debated. Sometimes passionate users create "sockpuppets" (fake user accounts) to advance their cause. Wikipedia has created a formal dispute resolution process and has even had to lock down some of the more controversial topics to stop continuous online combat.

A professor at George Mason University in Virginia required his students to create a credible Internet hoax as a class assignment. T. Kelly Mills teaches a course called "Lying About the Past," and his students have, at least temporarily, convinced the world that a woman opened a steamer trunk and found clues that her long-dead "Uncle" Joe was a serial killer. In other hoaxes, they faked up a beer recipe from 1812, complete with a webpage, beerof1812.com, and invented the tale of a pirate who roamed Chesapeake Bay in the 1870s. Mills has taught this course since 2008, and, as was observed in an *Atlantic* article, the Internet community is getting better at detecting online hoaxes.[371] The one that his students tried in 2012, about the serial killer, was debunked on reddit within twenty-six minutes.[372]

**Mince Your Metadata.**

The revelations of Edward Snowden brought the word "metadata" into the public's vocabulary. Until then, it was mainly a term for librarians to discuss at their conferences. Now we all know that when you make a phone call, the numbers involved and length of the call are captured, and that's a whole different issue from eavesdropping on the actual conversation. Many court cases have hinged on metadata, such as exactly when certain files were created, changed, or deleted.

In a similar fashion, your computer, smartphone, perhaps even your TV or cable box is tracking your activities. On February 1, 2004, when Janet Jackson's costume malfunctioned during the Super Bowl halftime show, digital video recorder maker TiVo was able to report that it was the most searched and replayed TV moment in history. This announcement revealed that the company is tracking viewing habits at a detailed level, and, indeed, we now know that it sells that information to advertisers.[373]

Here are some ways to make your metadata go away, or better yet, avoid generating it in the first place:

*Set your camera, smartphone, and other devices to "location off" mode.* Yes, it is handy to be able to find your way to the mall, or the Apple store in the mall, using smartphone navigation. In fact, people who have tried Google Glass say the navigation feature is its greatest asset though you might get a ticket for using it while driving.[374] One cheeky friend tweeted a photo of the stop sign as you leave the Google campus from his rental car, moments after receiving his "Google Glass Explorer's" version of the product.

For occasions where you actually want GPS functionality, it's no big deal to enable it temporarily. This is a true example of having to make a trade-off, because GPS (and other location capabilities such as the nearest Wi-Fi tower) provides one of the fastest-growing ways to track you with your own devices.

*If you have captured location data on a photo, remove it as soon as possible.*
There is really no need to be reminded that you took that photo in front of the Taj Mahal or the Grand Canyon, and, frankly, for selfies and twoshots in bars, you're probably best off forgetting the precise details. Find a piece of software that will "delete EXIF data." It's true that major services like Flickr and Facebook strip off this data before posting, but do you want *them* to have your exact location? People have been embarrassed, and worse, by emailing photos with GPS data intact.

Curious if a photo still has location data on it? If you don't mind uploading it to a website, www.exifdata.com will show you if there are GPS coordinates on an image, as well as a plethora of other information about the camera that took it.

*Alter your phone usage habits.*
Throughout *Technocreep*, I try to debunk the idea that if you have nothing to hide, you have nothing to worry about. Most people hold that belief at their peril. As just one example, the revelation that the National Security Agency captures almost 200 million SMS messages per day should give phone texters pause to consider.[375] Who has not, in a moment of anger, or cheekiness, sent a text that might send the wrong message? Something along the lines of "I'm going to kill you" or "I hope her plane crashes" or even some hilariously inappropriate message that was invented by the phone's attempt to correct your typing.

A fun website, www.damnyouautocorrect.com, contains priceless examples of autocorrection goofs like "Spent all day fisting with my dad," which was supposed to be "Spent all day fishing with my dad." Of course this might have happened because the phone's custom dictionary contained the word "fisting" and did not contain "fishing," but let's not go there.

Let's face it, if major government spy agencies want to track your phone calls, web surfing, or emails they will find a way to do it, perhaps with the help of an entity in another country to get around the "don't spy on our own citizens" rules. However, you can make yourself somewhat less visible and harder to track through the use of "burner cell phones," which are not tied directly to you (buy them with cash) and using soft phone services like Skype or a VOIP provider. Without wanting to endorse any of those, one does claim enhanced security if you use their service along with the Cisco secure VPN Router–007VoIP Special Edition. "Your browsing, Email correspondence and VoIP Telephony will not only never (be) intercepted any more," they say. "Even your internet provider (ISP) will not know what data you are transferring."[376]

That's probably overstating the case because of "deep packet inspection tools" and other sophisticated technologies. But there's no doubt that sometimes this kind "security through obscurity" can work in your favor. And sometimes not. There are rumors that spy agencies and law enforcement lavish special attention on those using techniques like TOR (The Onion Router) because they may have something to hide. Security guru Bruce Schneier explains how they do this on his blog.[377]

*Clear the history from GPS devices.*

Earlier in *Technocreep*, we saw an example of a car renter who incurred significant extra charges because the car's GPS tattled on him. That unit was hidden and he couldn't do anything about it. But you can control the GPS that you use, both in rental cars and ones that you own. If there's any reason you might not want to be pegged at a certain place, it's a simple matter to "erase all."

*Turn each new technology back on itself.*

As technology moves forward, the opportunities to turn it back on itself will also increase. For example, as discussed earlier in *Technocreep*, advertisers are now looking for ways to figure out where you are, either from your location on your smartphone or even from background noise like a stadium announcer.

Two can play at that game. Smartphone apps like Lexa's Fake GPS Location allow you to "teleport to any place in the world with two clicks!" It's handy if you want to check in on a location-based service like Foursquare from a place far from your actual location. As for the background sound, Spoofcard (mentioned earlier) has recently added a feature for its subscribers that allows them to make their call "sound like it's coming from a nightclub, or near traffic."

*Use a virtual private network.*

One of the biggest pieces of incidental data transmitted when you use the Internet is your IP address, which can provide a good idea of your location. For a few dollars a month (or even for free if you're willing to put up with *their* advertising), a number of companies will rent you IP addresses in other countries and route your traffic through them. This service, called a Virtual Private Network (VPN), also allows you to access services that are only available in certain parts of the world. Of course, with the proper legal documentation, VPN companies may be forced to cough up information on you such as your credit card number, so you probably don't want to give them anything genuine other than our friend I.P. Freely's prepaid credit card.

**Set Your Own Info-Traps.**

While the technology deck may seem to be stacked against the average person, there are powerful ways to fight back against the encroachment of technology into our lives.

*Vary your name.*

A simple place to start is how you tell companies your name. Savvy magazine subscribers have long used varying forms of their name or address to keep track of how their information is being sold to advertisers and other companies. It is true that postal standards are starting to restrict address creativity, but there are still twenty-six possibilities for your middle initial. And who says you're limited to just one initial?

The same principle can be transferred to the online world. Besides name variants, you can have fun with your salutation. Aeroplan.ca, the frequent flyer program associated with Air Canada, offers a fascinating array of titles to choose from when you sign up. Beyond the boring Mr./Mrs./Ms./Master, you can be Captain, Senator, Rabbi, Professor, Judge, and the ever-popular Père. Do try to avoid Doctor. I was once called to help a sick passenger in-flight and had to explain that I wasn't that kind of a doctor.

*Set traps in your email accounts.*

Many email providers, including Google's Gmail and Microsoft's Outlook.com, support the "plus convention." Anything you put after the plus sign in the name portion of your email address is ignored by the mailer, but carried along so you can track it. This will allow you to monitor the proliferation of spam, and figure out who the guilty parties are.

*Have your own surveillance cameras.*

Dash cameras have become widespread in many countries, and have become a "must have" on Russia's wild and woolly road system. They serve as a countermeasure against organized crime gangs that stage car accidents then try to extort money from innocent drivers on the spot.

Even in North America, people are grabbing footage of car crashes and other events and giving or selling it to media outlets. Police cars routinely have dash cameras. You should just assume that you're being photographed whenever you drive (or ride the bus, tram, subway, etc.) and behave accordingly.

*Record your life (just in case).*

Why not just record everything from a camera on your body? Cameras that do "lifelogging" have been around for years, but the problem is who really wants to relive their life moment by moment? Makers of the Narrative Clip are betting that people will want to "tell their life stories in photos" as their inexpensive and unobtrusive camera grabs a photo every thirty seconds.[378] It also stores up to four thousand pictures and they can be transferred to a computer. Unless you are a rock star, a spy, or a politician, you are likely to find that most of those images don't need to be transferred because you'll never want to look at them again.

Of course, if your body camera is stolen, or smashed by police in a riot, you'll lose what may be the most important pictures of the day. Cameras and smartphone apps are now springing up that will upload your photos or videos continuously to a server, so your vicious beating video might live on after you. Just remember to leave the password in your will as part of your "digital legacy."

The rise of "wearables" such as fitness wristbands, and even more advanced devices like glucose monitoring sensors in contact lenses, will make lifelogging the norm for most people.[379]

*Make your computer or smartphone your personal spy.*

"Phone home" anti-theft software can be remotely activated to track a lost or stolen device, make it take pictures at regular intervals, even howl an alarm. Some products to do this include LoJack, Find My Phone, Find My iPhone, and Prey (www.preyproject.com). Prey is open source and free for up to three devices. You might even pinpoint the precise Starbucks where your thief is slurping his latte while using your laptop. That program can also lock up your device tight as a drum, and make it display a "This is a stolen computer" message.

One clever user in Brazil, who could not find any company that would insure her electric bicycle, placed an inexpensive Android phone inside its tailbox, and used Prey to create her own anti-theft system.[380]

*Track your stolen computer through its attempt to access Gmail or Dropbox.*

If you have neglected to install anti-theft software you still might luck out when your stolen device connects to the Internet. Through services such as Gmail or Dropbox you may be able to see the latest IP address of the AWOL device and do some sleuthing. This could require help from law enforcement, but there have been success stories.

Another trick that has worked is to have your phone automatically upload photos you take, say to Facebook. Yes, this is a potential privacy risk for you, but it's an even bigger one for people who steal your phone and use it to take "selfies." In a delightful U.K. case, police are looking for a "brainless thief" who not only used a stolen phone to upload self photos but also "disclosed, via instant messaging, that he lives in south London."[381] News reports on this story carried a picture of the suspect, proving that guys who steal a phone from a twelve-year-old kid at a tube station have no reasonable expectation of privacy.

My favorite device recovery story is the tale of New York jazz trombonist Nadav Nirenberg, who left his iPhone in a livery cab. The new "owner" of the phone was using it to send messages to women using a dating app. Nirenberg sent him a message offering a date, and including a picture of a pretty girl. "When the culprit arrived at Nirenberg's Brooklyn apartment building with wine," reported the Associated Press, "the musician greeted him with a $20 U.S. bill while holding a hammer—just in case."[382]

**Protest Loudly but Carefully.**

What can you do when you catch a company selling your name without your permission, targeting you unfairly with ads, or treating you in some other creepy, shoddy fashion? First, you need to make sure you did not actually agree to whatever behavior is offending your sense of personal privacy. As shown in the terrific movie *Terms and Conditions May Apply,* in most cases you will find that you somehow consented to this use of your information, and that's the end of it. If you are sure you've been wronged, there are many options available:

*Send a polite but firm email, probably to a bot-monitored mailbox.*

You could start your complaint process by sending an email to the company's "customer service" (or whatever) department. If, as is often the case, you can't find that on their webpage, you could try sending the mail to abuse (at) whatever web domain is associated with the malefactor. Be warned that companies have often set up unmonitored mailboxes and auto responders and there's no guarantee you will actually reach a real person.

If they are using a free email service like Gmail or Outlook.com, and you provide enough evidence of nastiness, the provider might even yank their account, though of course real bad guys will simply move on to another one.

*Escalate to emailing an executive.*

Depending on how annoyed you are, you could find the contact information for executives at the company, and gripe to them. It may be a little early to call your lawyers, though, as the legal framework around business-to-business information selling is still evolving. The sticky part is proving that you have actually suffered harm from a company selling your information. In a 2013 decision in Delaware, U.S. District Judge Sue Robinson dismissed a class action lawsuit against Google even though they had sneakily bypassed cookie blocking software by exploiting a loophole in the Safari web browser.[383] The plaintiffs could not prove that they had suffered actual harm from the tracking cookies. Google did not get off scot-free, however. The company paid $17 million in civil penalties to thirty U.S. states, and promised not to be so evil in the future.

*Complain to government agencies and consumer watchdog groups such as the American Civil Liberties Union and the Electronic Frontier Foundation.*

As discussed earlier in *Technocreep*, a string of U.S. Federal Trade Commission rulings, decisions by Privacy Commissioners in Canada, and similar actions in other countries show that valid complaints can be pursued through this channel. The caveat is that it may take a long time, there may be costs in some cases, and there are no guarantees of success because the laws around privacy-related issues are still being written and interpreted.

*Go to the media, or make your own media.*

Most TV stations, and even some other media outlets, have a reporter assigned full or part time to the "consumer beat." They love David and Goliath stories of wronged consumers fighting back against bad treatment from impersonal corporations. Often, just having the reporter call a company is enough to sort things out.

Or you can take a more direct approach. Musician Dave Carroll put his 2009 luggage handling complaint against United Airlines to music as "United Breaks Guitars." The video has garnered over thirteen million views on YouTube.[384] He even turned it into a book, *United Breaks Guitars: The Power of One Voice in the Age of Social Media*. If you lack Carroll's musical talent, or the $150 he says the video cost to produce, you can just try holding up your story on cards in front of you while wearing a woeful look. That approach seemed pretty effective for the Occupy Wall Street protestors.

*Watch your back when you protest.*

The fact that companies are taking online consumer protests seriously is, by and large, a good thing. United Airlines, for example, reportedly uses Dave Carroll's video in its employee training. But sometimes you can hit so close to the nerve that a company bites back, which in America tends to mean "goes after your money."

In 2009, a woman from Utah bought some merchandise from an online merchant in Michigan. She was unhappy with the company's service, and wrote a critical review on RipoffReport.com. The merchant then threatened her with a $3,500 "fine" for violating an obscure non-defamation clause on its website. To avoid it, she had to take down the critical comments on the review site within 72 hours. She was told that failure to pay might damage her credit rating.[385]

However, RipoffReport.com says they "never remove reports." The only option they offer is their "VIP Arbitration Service" in which the complaint is reviewed by a private arbitrator, who might even be a retired judge. The catch? There's a $2,000 "filing fee" for this service.

This case turned into a wonderful demonstration of what has been called the "Streisand Effect," the result of that singer's 2003 attempt to suppress photos of her Malibu home, which generated even more publicity around them. The company that threatened to fine its customer is now the subject of negative reviews posted all over the Internet. News stories have been done on the plight of this harassed consumer. There are even outraged reports on RipoffReport.com from people who have never done business with the offending firm, hence never accepted their bizarre "terms of service," and who therefore cannot be "fined." All promise that they would never buy a thing from the company.

**Future-Proof Yourself, Your Technology, and Your Knowledge.** If the remarkable revelations of Edward Snowden and the astonishing private sector forays into creepy technology discussed in this book teach us anything, it's that technocreepiness will continue to change and expand on a daily basis. As I said at the start of the book, most technology is not what it seems. It is more than that, with wheels turning within wheels and systems interlocking in ways that most people don't even know exist.

This is both blessing and curse. We will never "fix" our security problems or "deal with" creepy technology. Even if we move to a cabin in Patagonia, there will be drones and satellites overhead. The best we can hope for is to have the right knowledge and mindset to understand the latest antics of the NSA, the GCHQ, and that big shopping mall down the street.

Since you probably have something else that's your full-time job, here are some ways to keep informed about technocreepiness, and benefit from the knowledge of others:

*Use a search engine to stay up to date.*

Yes, the major search engines do capture your information, sell it to advertisers, and even try to track you. Still, nothing beats a quick online search to figure out, for example, if you should download one of the programs mentioned in this book, in an article, or elsewhere. Thanks to search engine ranking, the first few hits (though not always the first one—there have been cases of search engine spoofing) will usually give you a good idea of whether something has a good or bad reputation.

*Check for scams.*

A simple but powerful trick is to put the word "scam" or "ripoff" into a search engine along with whatever you're interested in. As an illustration, a search on a major travel site, coupled with the word "scam," produced the warning that if you buy an airline ticket from a certain travel bidding site, "I can assure you that you will be the last one the airline rebooks. That's the price you can pay for buying a deeply discounted ticket."[386] It makes sense that an airline will take care of its best-paying customers first, but that information probably isn't displayed on any official policy page.

*Anticipate the next "DNA fingerprinting" type of breakthrough and prepare for it.*

A few decades ago, criminals were oblivious to leaving their DNA at crime scenes because there was no feasible way of identifying them. Even semen samples left by rapists were only identifiable to the blood group (A, B, AB, O) level. All that changed in the 1980s, when a technological advance (DNA sequencing) combined with legal changes to bring about the world we know today through television shows like *CSI: Crime Scene Investigation*. It behooves us to think now about what the next revolutionary technologies will be, and to understand that they won't just be used to hunt down criminals. We may all be targets.

If I had to put money on one thing that will be haunting people in even creepier ways in the very near future, it's their voluntarily shared photographs and videos. Teenagers who are posting X-rated shots of themselves today will be looking for jobs in a few years, and those images will be eminently findable. Facial recognition technology is moving so fast that the *Minority Report* scenario of being identified on the street, even in a crowd, will become reality. There is zero doubt in my mind that many computers, both governmental and private, are sucking up every digital image we share, just waiting for the day when it can be indexed and cross-referenced the way the LAPD did with tattoos back in 1959.

### Conspire with Like-Minded Folks and Participate in the Societal Dialogue.

It may seem odd to conclude with a call to conspiracy. However, I hope you now appreciate that you are already the target of many technology-fueled conspiracies, so it seems only right that you should have the tools to fight back.

*Hang with some hackers.*

My own introduction to the techno-conspiratorial mindset came decades ago, as I attended some of the New York City meetings of a group called 2600. This club took its name from the not-so-secret "operator mode" frequency (2600 Hz) that could unlock some of the wonders of the phone system. Their meetings were held at a midtown Manhattan office tower that had a large bank of public phones. Periodically during these informal gatherings, small groups would wander over to those pay phones to try out some trick they had just learned. I was pleased to see that NYC2600 is still meeting in roughly the same neighborhood, and has even managed to strike an ironic truce with Citigroup Center's building security folks, who seem to let them convene in peace in the building's food court to discuss ways to exploit flaws in security.[387]

Face-to-face hacker meetings may seem quaint in an age when ideas can be passed around the world in a heartbeat through the Internet. Still, they provide that extra creative spark. I highly recommend the DEF CON and Black Hat conferences as places to really understand those "wheels within wheels" that make technology so fascinating. I have had the privilege to speak at both events and enjoy them immensely.[388] [389] I always learn something. A lot in fact.

As a guide to groups that are doing deep thinking about technology issues, at least in North America, you could certainly start with the ones who show up at the DEF CON conference each year.[390]

One of the best things at DEF CON 2012 was that the exhibit floor organizers managed to intersperse the recruiting tables of government agencies (the NSA, CIA, FBI) in between privacy watchdog groups like the Electronic Frontier Foundation and the American Civil Liberties Union. Just sitting next to somebody different for four days undoubtedly did everyone a lot of good all around. Sadly, the government agencies were told they were not welcome (at least in their official suits) at the DEF CON 2013 conference, so this opportunity for human-to-human interaction was lost.

*Trade consumer information and reviews*.

Consumer conspiring is not limited to high tech issues, though the tools to accomplish it are often high tech. Coupon sharing sites and "where to find the best price for gasoline" webpages have popped up. Online travel bidding sites like Priceline have their own secret pricing formulas, so sites like biddingfortravel.com have arisen. On that one, you can find out things you are not really supposed to know, like which hotels in a city fall into which star categories. Customers also share winning and losing bid information, strategies, and even the secrets of the "free re-bids." It levels the playing field between consumer and company, though of course it creates a new divide between customers who have the insider information and those who don't.

While they can be the victim of bogus entries, there are now excellent review sites for travel, electronic products, cars, and almost anything you might want to investigate. Some even push the envelope a bit, such as when Flyertalk.com notifies its members about those ultra-cheap "mistake airfares" that airlines occasionally post and sometimes even honor for public relations reasons.

These techniques might not help you hide your identity or protect your personal information. But they are indicative of a mindset that says the consumer has the right to turn the tables on companies by using their own technological innovations.

*Accept the responsibility to stay informed, speak out, and vote on technocreepiness.*

With these tools and knowledge comes a responsibility to participate in the societal discussion of how much creepy technology we are willing to tolerate, for exactly which benefits.

Different societies will undoubtedly make different choices. In Estonia, for example, a single, mandatory, government-issued identity card is used to vote, pay taxes, shop, get medical care, and even ride public transit. Privacy advocates in the United States have staunchly opposed such a national ID card, fearing it could be misused to create a police state.

The revelations of Edward Snowden, Chelsea Manning, and others have revealed a previously hidden and creepy governmental addiction to surveillance and information hoarding.

The front pages of many science, technology, and medical journals have hints that creepier things are coming our way, at an increasing pace.

The maturing of the "digital natives," who have always known and accepted 24/7 connectivity, will alter our social discourse on what is creepy and what is cool.

One thing is certain. The hairs on the back of our collective neck are going to be working overtime for the foreseeable future.

You have been warned.

# Acknowledgments

I am extremely grateful to my family, friends, and colleagues who have supported me in the adventure of researching and writing this book.

My loving wife, Keri, and my delightful son, Jordan, have been there for me every single day with advice, energy, encouragement, references, criticism, and great ideas. Henry Mullish of New York University awakened my interest in technology way back in the 1960s. By allowing a teenage nerd to assist him in preparing some of his books, he also inspired a great love of the craft of writing.

My students and colleagues at the University of Calgary, and from twenty-five years of the Shad Valley Program, have all played a role in shaping the ideas behind this book. I'd especially like to thank Robert D. Acker, Paul Dickinson, Cullen Jennings, Zak Karbalai, Kingson Lim, Brian Lynch, Kathy Macdonald, David Moloney, Nate Dekens, Hervé St. Louis, and the robot at Google Alerts for sending me such amazing and creepy ideas and stories on a regular basis. They bring me great examples of technocreepiness, often with the same proud look the cat has when he carries in a half-eaten mouse.

Dr. Duncan Chappell and CBC producer Dave Redel played a seminal role in this project when we worked and played together in the 1980s creating *Crimes of the Future* for CBC IDEAS. Drs. Simone Fischer-Hübner and Penny Duquenoy have kindly involved me in the activities of the European privacy community through the FIDIS project. Countless stimulating presenters at the DEF CON, Black Hat, 2600, and Computers, Freedom, and Privacy conferences have given me terrific ideas, which are gratefully acknowledged. Dr. John Demartini, an inspired thinker about human behavior, gave me a framework that kept me from creeping myself out while researching this material.

I owe special thanks to John Oakes of OR Books for instantly believing in this project and for shepherding me through the journey of being a first-time book author. Staff members Justin Humphries, Courtney Andujar, Natasha Lewis, and Emily Freyer have also contributed greatly to this book.

Finally, I'd like to thank all my colleagues in the media. They asked many of the pesky, probing questions that led to this book. Even when they wanted me in the TV studio at 4 AM for a live interview, their provocation was exactly what I needed to stay sharp in this constantly changing field.

# Bibliography

Albrecht, Katherine, McIntyre, Liz. *Spychips: How Major Corporations and Government Plan to Track Your Every Purchase and Watch Your Every Move*. Nashville, TN: Nelson Current, 2005.

Angwin, Julia. *Dragnet Nation: A Quest for Privacy, Security, and Freedom in a World of Relentless Surveillance*. New York: Times Books, 2014.

Assange, Julian, *et al. Cypherpunks: Freedom and the Future of the Internet*. New York: OR Books, 2012.

Benjamin, Medea. *Drone Warfare: Killing by Remote Control*. New York: OR Books, 2012

Cavoukian, Ann, Tapscott, Don. *Who Knows: Safeguarding Your Privacy in a Networked World*. New York: McGraw Hill, 1997.

Davies, Kevin. *The $1,000 Genome: The Revolution in DNA Sequencing and the New Era of Personalized Medicine*. New York: Free Press, 2010.

Greenberg, Andy. *This Machine Kills Secrets: How WikiLeakers, Cypherpunks, and Hacktivists Aim to Free the World's Information*. New York: Penguin Group, 2012.

Greenwald, Glenn. *No Place to Hide: Edward Snowden, the NSA, and the U.S. Surveillance State*. New York: Metropolitan Books, 2014.

Kurzweil, Raymond. *The Singularity is Near: When Humans Transcend Biology*. New York: Viking Penguin, 2005.

Nissenbaum, Helen. *Privacy in Context: Technology, Policy, and the Integrity of Social Life*. Stanford, CA: Stanford University Press, 2010.

Schneier, Bruce. *Beyond Fear: Thinking Sensibly About Security in an Uncertain World*. New York: Copernicus, 2003.

Spar, Debora. *The Baby Business: How Money, Science and Politics Drive the Commerce of Conception*. Cambridge, MA: Harvard University Press, 2006.

Turkle, Sherry. *Alone Together: Why We Expect More from Technology and Less from Each Other*. New York: Basic Books, 2011.

Whitaker, Reg. *The End of Privacy: How Total Surveillance is Becoming a Reality*. New York: The New Press, 1999.

# References

All online resources accessed and verified in early 2014, but subject to change and deletion.


1. Blackwell, Tom. "Do-it-yourself brain stimulation has scientists worried as healthy people try to make their minds work better," *National Post*, June 12, 2013, accessed at http://news.nationalpost.com/2013/06/12/transcranial-direct-current-stimulation-tdcs-technology/
2. Clark, Liat. "Facebook is using AI to analyze the emotions behind your posts," *wired.co.uk*, September 23, 2013, accessed at http://www.wired.co.uk/news/archive/2013-09/23/facebook-deep-learning
3. Gannes, Liz. "Passwords on Your Skin and in Your Stomach: Inside Google's Wild Motorola Research Projects (Video)," *All Things D*, June 3, 2013, accessed at http://allthingsd.com/20130603/passwords-on-your-skin-and-in-your-stomach-inside-googles-wild-motorola-research-projects-video/
4. Eadicicco, Lisa. "Google Glass Will Track Your Gaze, Patent Hints," *Laptop*, August 14, 2013, accessed at http://blog.laptopmag.com/google-glass-patent, and describing U.S. Patent 8,510,166, issued August 13, 2013 and assigned to Google, Inc
5. BBC. "Tesco petrol stations use face-scan tech to target ads," November 4, 2013, accessed at http://www.bbc.com/news/technology-24803378
6. Hood, Leroy. Personal communication, Chicago, IL, February 14, 2014.
7. Carr, David. "Giving Viewers What They Want," *New York Times*, February 24, 2013, accessed at http://www.nytimes.com/2013/02/25/business/media/for-house-of-cards-using-big-data-to-guarantee-its-popularity.html
8. Gymrek, Melissa, *et al.* "Identifying Personal Genomes by Surname Inference," *Science*, January 18, 2013, 339(6117) p. 321-324, accessed at DOI: 10.1126/science.1229566.
9. Dobbs, Sarah. "50 Genuinely Creepy Horror Movies," Den of Geek, February 22, 2014, accessed at http://www.denofgeek.us/movies/horror-movies/22381/50-genuinely-creepy-horror-movies
10. http://www.reddit.com/r/Askreddit/comments/1d2v7i/parents_of_reddit_what_is_the_creepiest_thing/
11. Gravitz, Lauren. "When Your Diet Needs a Band-Aid," *MIT Technology Review*, May 1, 2009, accessed at http://www.technologyreview.com/news/413323/when-your-diet-needs-a-band-aid/
12. Seltzer, Leslie J., *et al.* "Social vocalizations can release oxytocin in humans," September 7, 2010, *Proceedings of the Royal Society of Biological Sciences*, 277(1694) pp. 2661-2666.
13. Wallace, Richard S. "From Eliza to A.L.I.C.E.," accessed at www.alicebot.org/articles/wallace/eliza.html
14. Foner, Leonard, N. In Dautenhahn, Kirstin, ed., *Human Cognition and Social Agent Technology*. Amsterdam: John Benjamins Publishing, 2000, p. 326.
15. Forums of Loathing. Accessed at forums.kingdomofloathing.com/vb/showthread.php?t=97928.J
16. This demonstrates how Hank's knowledge base was populated with information to help the Coca-Cola company deal with common rumors, such as being owned by the Mormon Church. See http://www.snopes.com/cokelore/mormon.asp. It also illustrates the limits of Hank's machine intelligence since he was asked about his own theological beliefs but was triggered by the word "Mormon" to reply about stock ownership
17. http://coca-cola-corporate.com.yeslab.org/contactus/
18. University of Reno. "Interactive Catholic Confessional," accessed at http://www.unr.

edu/art/delappe/first%20works/Confessional/Confessional%20MAIN.html
19. Wade, Peter. "For Yom Kippur, Synagogue Invites Congregation to Tweet Their Sins," *Fast Company*, September 13, 2013, accessed at http://www.fastcompany.com/3017468/fast-feed/for-yom-kippur-synagogue-invites-congregation-to-tweet-their-sins
20. Warren, Tom. "This is Cortana, Microsoft's answer to Siri," *The Verge*, March 3, 2014, accessed at http://www.theverge.com/2014/3/3/5465264/microsoft-cortana-windows-phone-screenshots
21. Kurzweil, Raymond. *The Singularity is Near: When Humans Transcend Biology*. New York: Viking Penguin, 2005.
22. Lovelace, Lady Ada. Quoted in Epstein, *et al.*, *Parsing the Turing Test*. New York: Springer Science, 2009, p. 53.
23. Freud, Sigmund. "Das Unheimlich," Imago, Bd. V., 1919; reprinted in *Sammlung, Fünfte Folge*. Translated by Alix Strachey, accessed at http://web.mit.edu/allanmc/www/freud1.pdf
24. Mori, Mashahiro. "Bukimi no Tani–The Uncanny Valley," translated by K.F. MacDorman & T. Minato. *Energy*, 7(4) p. 33–35, accessed at http://www.android-science.com/theuncannyvalley/proceedings2005/uncannyvalley.html
25. Harmon, Amy. "Making Friends With a Robot Named Bina48," *New York Times*, July 4, 2010, accessed at www.nytimes.com/2010/07/05/science/05robotside.html
26. FBI Special Agent Daniel R. Geneck. Criminal Complaint, dated April 21, 2013, accessed http://www.justice.gov/usao/ma/news/2013/April/criminalcomplaint1304211847.pdf
27. Klontz, Joshua C., Jain, Anil, K. "A Case Study on Unconstrained Facial Recognition Using the Boston Marathon Bombings Suspects," Technical Report MSU-CSE-13-4, accessed at http://www.cse.msu.edu/biometrics/Publications/Face/KlontzJain_CaseStudyUnconstrainedFacialRecognition_BostonMarathon BombimgSuspects.pdf
28. Schick, Shane (interviewer). "Security Lessons learned from the Boston tragedy (video interview)," CIO Association of Canada, accessed at http://www.youtube.com/watch?v=k5EK-tZlYvo
29. Wadhwa, Tarun. "The Next Privacy Battle: Cameras That Judge Your Every Move," *Forbes*, August 30, 2012.
30. Iwamoto, Kentaro, *et al.*, "Cigarette smoke detection from captured image sequences," in *Image Processing: Machine Vision Applications III (2010)*, accessed at http://www.sip.tuat.ac.jp/~tanaka/pdf/cigarette.pdf
31. New York Civil Liberties Union. "Who's Watching? Video Camera Surveillance in New York City and the Need for Public Oversight," Fall 2006, accessed at http://www.nyclu.org/pdfs/surveillance_cams_report_121306.pdf
32. This camera has now moved to 47th and Broadway. See it, and thousands of other cameras, live at http://www.earthcam.com/usa/newyork/timessquare/?cam=tsstreet
33. Hughes, Mark. "CCTV in the spotlight: one crime solved for every 1,000 cameras," *The Independent*, August 25, 2009, accessed at http://www.independent.co.uk/news/uk/crime/cctv-in-the-spotlight-one-crime-solved-for-every-1000-cameras-1776774.html
34. BBC. "Six crimes a day solved by CCTV, Met says," accessed at http://www.bbc.co.uk/news/uk-england-london-12080487
35. Lorenc, Theo, *et al.*, "Environmental interventions to reduce fear of crime: systematic review of effectiveness," Systematic Reviews 2(30), 2013, accessed November 24, 2013 at www.systematicreviewsjournal.com/content/2/1/30
36. Schneier, Bruce. *Beyond Fear: Thinking Sensibly About Security in an Uncertain World*. Copernicus: New York, 2010.
37. Hui, Stephen. "ICBC offers facial-recognition technology to Vancouver police's riot investigation," *Straight.com*, June 17, 2011, accessed at http://www.straight.com/news/icbc-offers-facial-recognition-technology-vancouver-polices-riot-investigation
38. Canadian Press. "Use of ICBC data okayed to identify rioters," June 22, 2011, accessed at http://www.cbc.ca/news/canada/british-columbia/use-of-icbc-data-okayed-to-identify-rioters-1.1064189
39. Vancouver Police Department."IRIT recommends charges against 350th suspected rioter, July 23, 2013, accessed at http://mediareleases.vpd.ca/2013/07/23/irit-recommends-charges-against-350th-suspected-rioter/

40. See http://www.gigapixel.com/image/gigapan-canucks-g7.html
41. Savage, Charlie. "Facial Scanning Is Making Gains in Surveillance," *New York Times,* August 21, 2013, accessed at http://www.nytimes.com/2013/08/21/us/facial-scanning-is-making-gains-in-surveillance.html
42. Merchant, Brian. "We're 'Five Years Off' From Homeland Security Using Facial Recognition to Profile U.S. Citizens," *Motherboard*, August 21, 2013, accessed at http://motherboard.vice.com/blog/were-five-years-off-from-homeland-security-using-facial-recognition-to-profile-us-citizens
43. Langfitt, Frank. "In China, Beware: A Camera May Be Watching You," National Public Radio, January 29, 2013, accessed at http://www.npr.org/2013/01/29/170469038/in-china-beware-a-camera-may-be-watching-you
44. Keenan, Jordan. "The End of Anonymity and the Quest for Better Gadgets," *Motherboard.tv,* accessed at http://motherboard.vice.com/blog/the-end-of-anonymity-and-the-quest-for-better-gadgets
45. Crowson, Scott. "Police hunt for mom in B.C.," *Calgary Herald*, August 18, 2001, p. A1.
46. Weber, Peter. "Watch why Russians put dashboard cameras in their cars," *The Week,* September 25, 2013, accessed at http://theweek.com/article/index/250148/watch-why-russians-put-dashboard-cameras-in-their-cars
47. Shachtman, Noah. "Pentagon Kills LifeLog Project," *Wired,* February 4, 2004, accessed at http://www.wired.com/politics/security/news/2004/02/62158
48. Kallstrom, Martin. "How Lifelogging is Transforming the Way We Remember, Track Our Lives," *Wired*, June 10, 2013, accessed at http://www.wired.com/insights/2013/06/how-lifelogging-is-transforming-the-way-we-remember-track-our-lives/
49. Barber, Nick. "Exclusive: MIT's 'Kinect of the Future' looks through walls with X-ray like vision," *IDG News Service*, October 11, 2013, accessed at http://www.youtube.com/watch?v=caVZZYbZV_4
50. O'Brien, Terrence. "Creepy new Air Force camera can identify and track you from far, far away," *Engadget*, May 20, 2011, accessed at http://www.engadget.com/2011/05/20/creepy-new-air-force-camera-can-identify-and-track-you-from-far/
51. Photon-X. Demo video, accessed at http://www.photon-x.com/flash/robinspin.html
52. As explained in Steven J. Levy's book, *Hackers: Heroes of the Computer Revolution,* the term "hacker" was originally an honorific reserved for skilled programmers who could "hack away" at virtually any problem. However, the word has largely been re-purposed to mean someone who uses technology in a malevolent way. In this book, both senses are implied at various points.
53. Rastreador de Namorado, app (in Portuguese), accessed at http://rastreadordenamorado.com.br/
54. Laird, Lorelei, "Victims are taking on 'revenge porn' websites for posting photos they didn't consent to," *ABA Journal*, November 1, 2013, accessed at http://www.abajournal.com/magazine/article/victims_are_taking_on_revenge_porn_websites_for_posting_photos_they_didnt_c/
55. Federal Bureau of Investigation. Press release, September 26, 2013, accessed at http://www.fbi.gov/losangeles/press-releases/2013/temecula-student-arrested-in-sextortion-case-involving-multiple-victims
56. U.S. Federal Trade Commission. "FTC Approves Final Order Settling Charges Against Software and Rent-to-Own Companies Accused of Computer Spying," April 15, 2013, accessed at www.ftc.gov/opa/2013/04/designerware.shtm
57. Neil, Martha. "7 Retailers Settle with FTC, Agree to Stop Spying on Up to 400,000 Computer Rental Customers," Sept 26, 2012, accessed at http://www.abajournal.com/news/article/7_Companies_Settle_With_FTC_Agree_to_Stop_Using_Rental_Computer_Webcams_to/
58. Ilgrenfritz, Richard. "LMSD reaches agreement in Webcam case," *Mainline Media News*, October 13, 2010, accessed at http://www.mainlinemedianews.com/articles/2010/10/13/main_line_times/news/doc4cb3a75f46e9f675801205.txt
59. Accessed at http://www.keralacm.gov.in/
60. Accessed at http://cam.f-arts.co.jp
61. Young, Nora. "Always-on Video Portal with Danny Robinson (Full interview)," CBC Radio's Spark, September 13, 2013, accessed at http://www.cbc.ca/player/Radio/Spark/Extended+Interviews/ID/2406044767/

62. iStrategyLabs. "Take an Instant Selfie With This Magical Mirror," blog posting, April 7, 2014, accessed at http://istrategylabs.com/2014/04/take-an-instant-selfie-with-this-magical-mirror/
63. Farrar, W., Ariel, B. "Self-Awareness to Being Watched and Socially-Desirable Behavior: A Field Experiment on the Effect of Body-Worn Cameras on Police Use-of-Force," accessed at http://www.policefoundation.org/sites/g/files/g798246/f/201303/The%20Effect%20of%20Body-Worn%20Cameras%20on%20Police%20Use-of-Force.pdf
64. Stross, Randall. "Wearing a Badge, and a Video Camera," *New York Times*, April 6, 2013, accessed at http://www.nytimes.com/2013/04/07/business/wearable-video-cameras-for-police-officers.html?_r=2&
65. Vancouver Police Department. "2011 Stanley Cup Riot Review," September 6, 2011, accessed at http://vancouver.ca/files/cov/2011-stanley-cup-riot-VPD.pdf
66. Lee, Hyunho. "Korean researchers demonstrate a new class of transparent, stretchable electrodes," *Asia Research News*, May 30, 2013, accessed at http://www.researchsea.com/html/article.php/aid/7724/cid/2/research/korean_researchers_demonstrate_a_new_class_of_transparent__stretchable_electrodes.html
67. Accessed at http://www.prserve.com/sunshade/
68. Barrett, Chris. "Google Glass - The First Fight & Arrest Caught on Glass - July 4 Wildwood, NJ boardwalk," July 14, 2013, accessed at https://www.youtube.com/watch?v=4isOSntnpo8
69. Google. Glass (etiquette guidelines). n.d., accessed at https://sites.google.com/site/glasscomms/glass-explorers
70. Rosenblum, Andrew. "Spy vs. Spy: Casinos Can't See The Cameras Hidden Up Gamblers' Sleeves," *Popular Science*, August 8, 2011, accessed at http://www.popsci.com/technology/article/2011-06/spy-vs-spy-casinos-cant-see-cameras-hidden-gamblers-sleeves
71. MiKandi. "First-Ever Google Glass Porn," accessed at http://www.youtube.com/watch?v=Xxt24JoLlPE
72. *Los Angeles Times*. "Burger King does not want its employees to take baths," August 13, 2008, accessed at http://opinion.latimes.com/opinionla/2008/08/burger-king-doe.html
73. Sachs, Wendy. "The secret life of my sixth grader," CNN, November 27, 2012, accessed at http://edition.cnn.com/2012/11/27/living/child-social-media
74. The National Campaign, "Sex and Tech," 2008, accessed at http://www.thenationalcampaign.org/
75. Bulatovic, Peja. "Nudity filter helps Chatroulette clean up," January 20, 2011, at http://www.cbc.ca/news/technology/story/2011/01/20/algorithm-tech-nudity-chatroulette-filter-flashers.html
76. Badabing!, iTunes store, accessed at https://itunes.apple.com/us/app/badabing!/id548536602?mt=8
77. SnapChat Save Pics, March 2014, accessed at https://play.google.com/store/apps/details?id=com.appztastic.snapchatsavepics
78. Acquisti, A., Gross, R., Stutzman, F. "Faces of Facebook: Or, How The Largest Real ID Database In The World Came To Be," research paper accessed at http://www.heinz.cmu.edu/~acquisti/face-recognition-study-FAQ/
79. Walters, Helen. "The battle between public and private: Alessandro Acquisti at TEDGlobal 2013," accessed at http://blog.ted.com/2013/06/14/the-battle-between-public-and-private-alessandro-acquisti-at-tedglobal-2013
80. Vale, Jack. "Social Media Experiment," November 18, 2013, accessed at http://www.youtube.com/watch?feature=player_embedded&v=5P_0s1TYpJU
81. Belgian Financial Sector Federation, "Amazing mind reader reveals his 'gift," accessed at http://www.youtube.com/watch?v=F7pYHN9iC9I
82. Kim, Susanna. "Tenn. Family Sues After Seeing Altered, 'Offensive' Images of Son with Down Syndrome," ABC News, April 29, 2013, accessed at http://abcnews.go.com/Business/tenn-family-sues-alter-images-son-syndrome/story?id=19050815#.UYFcHUqNAR8
83. Oved, Macro Chown. "Stolen Instagram baby photos used for sexual role play," *Toronto Star*, May 1, 2013, accessed at http://www.thestar.com/life/technology/2013/05/01/stolen_instagram_baby_photos_used_for_sexual_role_play.html
84. *Ibid*.
85. Gruchawka, Steve. "Using the Deep Web," techdeepweb.com, n.d., accessed at http://techdeepweb.com,

86. Paglieri, Jose. "The Deep Web you don't know about," CNN Money, March 10, 2014, accessed at http://money.cnn.com/2014/03/10/technology/deep-web/
87. Hern, Alex. "Silk Road 2.0 resurrects online drugs marketplace," *The Guardian,* November 7, 2013, accessed at http://www.theguardian.com/technology/2013/nov/07/silk-road-20-resurrects-online-drugs-marketplace
88. Claburn, Thomas. "Facebook Says User Data Is Price of Admission," *Information Week Government,* August 29, 2013, accessed at http://www.informationweek.com/applications/facebook-says-user-data-is-price-of-admission/d/d-id/1111349
89. Keenan, Thomas P. "On the Internet, Things Never Ever Go Away Completely," in *The Future of Identity in the Information Society IFIP—The International Federation for Information Processing* Volume 262, Springer Business Media, 2008, pp. 37-50.
90. Hill, Kashmir. "Payment Providers And Google Will Kill The Mug-Shot Extortion Industry Faster Than Lawmakers Can," *Forbes,* October 7, 2013, accessed at http://www.forbes.com/sites/kashmirhill/2013/10/07/payment-providers-and-google-will-kill-the-mug-shot-extortion-industry-faster-than-lawmakers/
91. Franklin County Ohio Clerk of Courts of the Common Pleas. Exhibit E, Affidavit of David London, June 27, 2012, accessed at http://www.leader.com/docs/AFFIDAVIT-OF-DAVID-LONDON-EXHIBIT-D-Defendants-Motion-to-Enforce-Settlement-27-Jun-2012-CLERK-COPY.pdf
92. Robertson, Adi. "Facebook deletes European facial recognition data, satisfying German privacy agency," *The Verge,* February 7, 2013, accessed at http://www.theverge.com/2013/2/7/3964550/facebook-deletes-european-facial-recognition-data
93. Butcher, Mike. "Facebook Turns Photo Tag Suggestions Back On In The US—Will Users Like It This Time?" *TechCrunch,* February 1, 2013, accessed at http://techcrunch.com/2013/02/01/facebook-turns-photo-tag-suggestions-back-on-in-the-us-will-users-like-it-this-time/
94. Form S-1 Registration Statement. U.S. Securities and Exchange Commission, accessed at http://www.sec.gov/Archives/edgar/data/1326801/000119312512034517/d287954ds1.htm
95. Taigman, Yaniv, *et al.*, "DeepFace: Closing the Gap to Human-Level Performance in Face Verification," prepared for conference on Computer Vision and Pattern Recognition (CVPR), June 2014, accessed at https://www.facebook.com/publications/546316888800776/
96. University of Massachusetts. "Labeled Faces in the Wild." Accessed at http://vis-www.cs.umass.edu/lfw/
97. Grandoni, Dino. "Facebook's New 'DeepFace' Program Is Just As Creepy As It Sounds," The Huffington Post, March 18, 2014, accessed at http://www.huffingtonpost.com/2014/03/18/facebook-deepface-facial-recognition_n_4985925.html
98. Polacek, Jeremy. "Facebook's Freaky DeepFace Program Knows Your Friends Better Than You Do," www.policymic.com, accessed at http://www.policymic.com/articles/85719/facebook-s-freaky-deepface-program-knows-your-friends-better-than-you-do
99. The Onion. "CIA's 'Facebook' Program Dramatically Cuts Agency's Costs," accessed at http://www.theonion.com/video/cias-facebook-program-dramatically-cut-agencys-cos,19753/
100. gregorrohrig. "Does what happens in the Facebook stay in the Facebook?," YouTube, May 29, 2007, accessed at http://www.youtube.com/watch?v=wogtTQs8Kzw
101. Turkle, Sherry, *Alone Together: Why We Expect More from Technology.* New York: Basic Books, 2011, p. 13.
102. Clayman, Margie, "Why do we post the things we do to the online world?," April 20, 2013, accessed at http://www.margieclayman.com/why-do-we-post-the-things-we-do-to-the-online-world
103. Harvey, Adam. "Camouflage from face detection," accessed at http://cvdazzle.com/
104. Prokop, Petr. Face Dazzler (app). accessed at https://play.google.com/store/apps/details?id=cz.zweistein.android.facedazzler
105. Terry, Dean. "EnemyGraph Facebook Application," blog posting, February 21, 2012, accessed at www.deanterry.com/post/18034665418/enemygraph
106. Broderick, Ryan, and Grinberg, Emanuella, "10 people who learned social media can get you fired," *CNN,* June 6, 2013, accessed at http://edition.cnn.com/2013/06/06/living/buzzfeed-social-media-fired/
107. MacMillan, Robert. "Model Linux Geek an MS User Too?," *Wired.com,* January 21,

2003, accessed at http://www.wired.com/software/coolapps/news/2003/01/57307
108. Keenan, Tom. "Mysterious poster 'geek chick' is Calgarian!" *Business Edge News Magazine*, January 30, 2003, accessed at http://www.businessedge.ca/archives/article.cfm/mysterious-poster-geek-chick-is-calgarian-2415
109. http://www.microsoft.com/nz/vstudio/productinfo/default.aspx
110. Moskvitch, Katia. "Internet of things: Should you worry if your jeans go smart?" *BBC News*, September 22, 2011, accessed at http://www.bbc.co.uk/news/business-15004063
111. Kelly, Heather. "The CNN 10: The future of driving," February 28, 2014, accessed at http://edition.cnn.com/interactive/2014/02/tech/cnn10-future-of-driving/
112. Brownlee, John. "GPS chips are now smaller than a match head," BoingBoing, February 12, 2009 at BoingBoing, accessed at http://gadgets.boingboing.net/2009/02/12/gps-chips-are-now-sm.html
113. *The Economist*. "Roads less travelled," The Economist, October 19, 2013, accessed at http://www.economist.com/news/united-states/21588097-oregon-wants-tax-motorists-miles-driven-not-petrol-burned-will-it-work-roads-less
114. CBC News."Car's black box convicts dangerous driver," April 14, 2004, accessed at http://www.cbc.ca/news/canada/car-s-black-box-convicts-dangerous-driver-1.469369
115. Yerak, Becky. "Motorists tap brakes on installing insurers' data devices," *Chicago Tribune,* September 20, 2013, accessed at http://articles.chicagotribune.com/2013-09-15/classified/ct-biz-0915–telematics-insure-20130915_1_insurance-companies-insurance-telematics-progressive-snapshot
116. Taylor, Alexis. "Ohio Father Cleared of Murder Thanks to Progressive Insurance 'Snapshot' Device," *Afro.com*, August 2, 2013, accessed at http://www.afro.com/sections/news/afro_briefs/story.htm?storyid=79320
117. Yerak, Becky. "Allstate to expand Drivewise to help parents keep tabs on teens," *Chicago Tribune,* May 29, 2013, accessed at http://articles.chicagotribune.com/2013-05-29/classified/chi-allstate-teen-drivers-20130529_1_teens-tool-allstate-corp
118. Lubinsky, Daryl. "GPS Tracking the Right Way," *Auto Rental News*, September/October 2011, accessed at http://www.autorentalnews.com/article/story/2011/09/gps-tracking-the-right-way-disclosure-fees-and-recovery.aspx
119. Ho, Janie. "GPS Keeping Tabs On Car Rentals," *CBS Evening News*, March 6, 2004, accessed at http://www.cbsnews.com/news/gps-keeping-tabs-on-car-rentals
120. Mehes, Adam. "Acme ordered to stop fining speeders," *Yale Daily News*, February 28, 2002, accessed at http://yaledailynews.com/blog/2002/02/28/acme-ordered-to-stop-fining-speeders/
121. CBC News. "Live traffic map uses Vancouver drivers' cellphone data," August 13, 2013, accessed at http://www.cbc.ca/news/canada/british-columbia/live-traffic-map-uses-vancouver-drivers-cellphone-data-1.1324932
122. BBC News. "Talking CCTV scolds offenders," BBC, April 4, 2007, accessed at http://news.bbc.co.uk/2/hi/6524495.stm
123. CNN, "Kendrick Johnson Death: Accident or Murder?," November 21, 2013, accessed at http://edition.cnn.com/video/data/2.0/video/crime/2013/11/22/ac-dnt-blackwell-kendrick-johnson-surveillance-tapes.cnn.html
124. Griffiths, Sarah. "Computer hackers can now hijack TOILETS: 'Smart toilet' users in Japan could become victim to Bluetooth bidet attacks and stealthy seat closing," *Mail Online*, August 5, 2013, accessed at http://www.dailymail.co.uk/sciencetech/article-2384826/Satis-smart-toilets-Japan-hacked-hijacked-remotely.html
125. Lomas, Natasha. "Wearable Tech For A Practical Problem: Spanish Startup Builds Alert System For Diaper Changing," *TechCrunch,* November 8, 2013, accessed at http://techcrunch.com/2013/11/08/siempresecos
126. Dick-Agnew, David. "IDEO's David Webster on Healthcare Design," *Azure Magazine*, September 23, 2013, accessed at http://www.azuremagazine.com/article/ideos-david-webster-on-healthcare-design/
127. "Ordering Pizza in the Future," accessed at http://www.youtube.com/watch?v=RNJl9EEcsoE
128. Estes, Adam Clarke. "Your Fuelband Knows When You're Having Sex," accessed at http://gizmodo.com/your-fuelband-knows-when-youre-having-sex-755620844
129. Ferenstein, Gregory. "How Health Trackers Could Reduce Sexual Infidelity," *TechCrunch*, July 5, 2013, accessed at http://techcrunch.com/2013/07/05/how-health-trackers-could-reduce-sexual-infidelity/
130. Sensing City. "What is Sensing City?" accessed at http://www.sensingcity.org/

131. Jha, Alok. "Tesco ends trial of CCTV spy chip on razor blades," *The Guardian*, August 22, 2003, accessed at http://www.theguardian.com/business/2003/aug/22/supermarkets.uknews
132. Burberry website. "RFID," accessed at http://ca.burberry.com/legal-cookies/privacy-policy/rfid/
133. *The Economist*. "Burberry goes digital; High-tech fashion," *The Economist*, 404.8803, September 22, 2012, p. 76.
134. Fung, Brian. "This snack-food corporation has a creepy plan to watch you in the grocery store," *Washington Post*, October 14, 2013, accessed at http://www.washingtonpost.com/blogs/the-switch/wp/2013/10/14/this-snack-food-corporation-has-a-creepy-plan-to-watch-you-in-the-grocery-store/
135. Samson, Ted. "U.S. senator demands suspension of phone-tracking system," *Infoworld*, November 28, 2011, accessed at http://www.infoworld.com/t/security/us-senator-demands-suspension-phone-tracking-system-180215
136. Renew London, "Renew release results of smartphones data capture," June 17, 2013, accessed at renewlondon.com/2013/06/renew-release-results-of-smartphone-data-capture/
137. Watson, Tom. "Official Statement on Renew Orb from CEO Kaveh Memari," posted August 12, 2013, accessed at renewlondon.com/2013/08/official-message-on-renew-orb-from-ceo-kaveh-memari
138. Renew London, "Presence ORB - Renew Technologies (R&D)," accessed at vimeo.com/6607410
139. Wilkinson, Glenn. "The Machines that Betrayed Their Masters," presentation at Black Hat Asia, Singapore, March 28, 2014.
140. Gittleson, Kim. "Data-stealing Snoopy drone unveiled at Black Hat," *BBC*, March 28, 2014, accessed at http://www.bbc.com/news/technology-26762198
141. WakeUpCallPage. "Walmart's RFID Chips," July 16, 2013, accessed at https://www.youtube.com/watch?v=XoNuKK4WtOY
142. United States Patent and Trademarks Office. "Identification and tracking of persons using RFID-tagged items in store environments," Patent #7,076,441, filed May 3, 2001, accessed at http://patft.uspto.gov
143. Yahoo! Finance. "VeriTeQ Partner Establishment Labs Receives CE Mark Approval to Market Motiva Implant Matrix Ergonomix Breast Implant with VeriTeQ's Q Inside Safety Technology," October 10, 2013, accessed at http;//finance.yahoo.com/news/veriteq-partner-establishment-labs-receives-123000807.html
144. Omnilink. "Gang Tracking and Intervention with Electronic Monitoring," accessed at http://www.omnilink.com/gang-activity-monitoring/
145. Information Week. "Apparel Maker Tags RFID For Kids' Pajamas," *Information Week*, July 15, 2005, accessed at http://www.informationweek.com/apparel-maker-tags-rfid-for-kids-pajamas/165702816
146. Obasanjo, Dave. "How Facebook Knows What You Looked at on Amazon," blog posting, February 17, 2014, accessed at http://www.25hoursaday.com/weblog/2014/02/17/HowFacebookKnowsWhatYouLookedAtOnAmazon.aspx
147. United States District Court, Northern District of California, San Jose Division. Case No. 5:13-md-02430-LHK , September 5, 2013, accessed at https://ia801904.us.archive.org/25/items/758246-160134104-google-motion-to-dismiss-061313/758246-160134104-google-motion-to-dismiss-061313.pdf
148. Davidson, Greg, *et al*. "The YouTube video recommendation system," RecSys '10 Proceedings of the fourth ACM conference on Recommender systems, pp. 293-296.
149. Snopes.com. Accessed at http://www.snopes.com/risque/kinky/panties.asp
150. ratboy. "Japanese Vending Machines 'Used Panties,'" YouTube, October 6, 2008, accessed at https://www.youtube.com/watch?v=iPqol5Qvq_E
151. Dugdale, Addy. "Cell-Phone Tech Uses Accelerometers to Spy on Employees," fastcompany.com, March 8, 2010, accessed at http://www.fastcompany.com/1575213/cell-phone-tech-uses-accelerometer-spy-employees
152. http://www.scenetap.com/faq
153. Etherinton, Darrell. "Secret Handshake Lets You Pay With Hand Gestures And Leap Motion–No Phone Or Card Required," *TechCrunch*, September 8, 2013, accessed at http://techcrunch.com/2013/09/08/secret-handshake-lets-you-pay-with-hand-gestures-and-leap-motion-no-phone-or-card-required/
154. Khambadkar, Vinitha, Folmer, Eelke. "GIST: a Gestural Interface for Remote Nonvisual Spatial Perception," 2013, accessed at http://eelke.com/files/pubs/gist.pdf
155. Stuff.tv. "Play the drums with your heartbeat," February 8, 2013, accessed at http://www.stuff.tv/play-drums-your-heartbeat/news

156. Oculus Joins Facebook. March 25, 2014, accessed at http://www.oculusvr.com/blog/oculus-joins-facebook/
157. Benjamin, Medea. *Drone Warfare: Killing by Remote Control*. New York: OR Books, 2012, p. 27.
158. McClure, Matt. "U of C researchers find lost heat costs Calgary homeowners millions," *Calgary Herald*, November 11, 2013, accessed at http://www.calgaryherald.com/researchers+find+lost+heat+costs+Calgary+homeowners+millions/9151052/story.html
159. Dr. Geoffrey Hay, Private Communication, September 14, 2013.
160. Kyllo v. United States, 533 U.S. (2001).
161. Goodin, Dan. "DIY aerial drone monitors Wi-Fi, GSM networks," *The Register*, August 5, 2011, accessed November 24, 2013 at www.theregister.co.uk/2011/08/05/flying_spy_drone/
162. Ullrich, Peter and Wollinger, Gina Rosa. "A surveillance studies perspective on protest policing: the case of video surveillance of demonstrations in Germany," *Interface*, (1):12-38 (May 2011), accessed at http://www.interfacejournal.net/wordpress/wp-content/uploads/2011/05/Interface-3-1-Ullrich-and-Wollinger.pdf
163. Duhigg, Charles. "How Companies Learn Your Secrets," *New York Times*, February 16, 2012, accessed at http://www.nytimes.com/2012/02/19/magazine/shopping-habits.html
164. Bode Technology. "What is Touch DNA?," accessed at http://www.bodetech.com/forensic-solutions/dna-technologies/touch-dna/
165. http://www.alliancedata.com/
166. Ryan, Tom. "Safeway: No Loyalty Card, No Discounts," *Retailwire.com*, April 3, 2012, accessed at http://www.retailwire.com/discussion/15920/safeway-no-loyalty-card-no-discounts
167. *The Economist*. "Robot recruiters," *The Economist*, April 6, 2013, accessed at http://www.economist.com/news/business/21575820-how-software-helps-firms-hire-workers-more-efficiently-robot-recruiters.
168. CBC News. "Depressed woman loses benefits over Facebook photos," November 19, 2009, accessed at http://www.cbc.ca/news/canada/montreal/depressed-woman-loses-benefits-over-facebook-photos-1.861843
169. Katzman, Alan. "Why Ivy League Admissions Officers Have No Choice But To Google College Applicants," *Business Insider*, accessed at http://www.businessinsider.com/why-ivy-league-admissions-will-google-you-2013-9
170. Mashable. "Oddball LinkedIn Endorsements to Give Your Connections," *Mashable.com*, n.d., accessed at mashable.com/2013/03/04/linkedin-endorsements-weird/#gallery/oddball-linkedin-endorsements-to-give-your-connections/520c1f723182972b420002a
171. Advertising Age. "Smell-Vertising Hits U.K. With Potato-Scented Bus Shelters," February 7, 2012, accessed at http://adage.com/article/creativity-pick-of-the-day/smell-vertising-hits-u-k-potato-scented-bus-shelters/232586/
172. Cofer, Jim. "RETRO TECH: Aromance Aroma Disc Player," accessed at http://jim-cofer.com/personal/2012/08/17/retro-tech-aromance-aroma-disc-player/
173. Bevers, Sabine. "10 Weird Sensory Marketing Tricks Companies Use On Us," March 13, 2013, at http://listverse.com/2013/03/13/10-weird-sensory-marketing-tricks-companies-use-on-us/
174. Aromasys. Accessed at http://www.aromasys.com
175. SerenaScent. Accessed at http://www.serenascent.com.au/index.php
176. United States Patent and Trademarks Office. "Scent-Emitting Systems," Patent #4,603,030, filed September 20, 1984, accessed at http://patft.uspto.gov
177. Oliver, Gabriel. "The 5 Most Unsettling Disney Theme Park Easter Eggs," cracked.com, August 14, 2012, at http://www.cracked.com/article_19977_the-5-most-unsettling-disney-theme-park-easter-eggs.html#ixzz2iOQJOSC5
178. Albrecht. Leslie. "Barclays Center's 'Signature Scent' Tickles Noses, Curiosity," *DNA Info New York*, May 20, 2013, accessed at http://www.dnainfo.com/new-york/20130520/prospect-heights/barclays-centers-signature-scent-tickles-noses-curiosity
179. Volpicelli, Gian. "The Multi-Sensory Internet Brings Smell, Taste, and Touch to the Web," *Motherboard*, November 10, 2013, accessed at http://motherboard.vice.com/blog/the-multi-sensory-internet-brings-smell-taste-and-touch-to-the-web
180. Bloxham, Andy. "Ikea layout 'intended to confuse shoppers'," *The Telegraph*, January 24, 2011, accessed at http://www.telegraph.co.uk/news/uknews/8278010/

Ikea-layout-intended-to-confuse-shoppers.html
181. PremEx. Blog posting, *FlyerTalk,* accessed at http://www.flyertalk.com/forum/3380935-post9.html
182. Creative Brand Communications. "Three Ways to Use Taste In Your Next Marketing Promotion," accessed at http://www.creative-brand.com/expertise/financial-brand-strategy-newsletters/three-ways-to-use-taste-in-your-next-marketing-promotion
183. Moss, Michael. "That Nacho Dorito Taste," *New York Times,* October 1, 2013, accessed at http://www.nytimes.com/2013/10/02/dining/the-nacho-dorito.html
184. CNN. "Doritos inventor to be buried with the snack that made him famous," September 26, 2011, accessed at http://eatocracy.cnn.com/2011/09/26/doritos-inventor-to-be-buried-with-the-snack-that-made-him-famous/
185. CTV Kitchener. "Bowland, Boone found guilty of not disclosing HIV," December 19, 2012, accessed at http://kitchener.ctvnews.ca/bowland-boone-found-guilty-of-not-disclosing-hiv-1.1086178
186. Xinhua. "Real-name AIDS testing creates controversy," June 3, 2013, accessed at http://usa.chinadaily.com.cn/china/2013-06/03/content_16562336.htm
187. Vonn, Michael. "British Columbia's 'seek and treat' strategy: a cautionary tale on privacy rights and informed consent for HIV testing," May 2012, accessed at http://www.aidslaw.ca/publications/interfaces/downloadFile.php?ref=2026
188. HuntSocCanada. "Do You Really Want to Know? 30 Sec PSA Ad," accessed at http://www.youtube.com/watch?v=AdqolAiKRdY
189. Office of the Privacy Commissioner of Canada. "Genetic Information, the Life and Health Insurance Industry and the Protection of Personal Information: Framing the Debate," November 2012, accessed at http://www.priv.gc.ca/information/research-recherche/2012/gi_intro_e.asp
190. Medline Plus. "Huntington disease," May 28, 2013, accessed at http://www.nlm.nih.gov/medlineplus/ency/article/000770.htm
191. Walker, Francis O. "Huntington Disease." *The Lancet*, 369(9557), pp. 218–228.
192. Dvorsky, George. "Nope. Scientists did not just find an 'alcoholism gene", November 28, 2013, accessed at http://io9.com/nope-scientists-did-not-just-find-an-alcoholism-gene-1473186030
193. The Rush. "Director John Zaritsky on his new documentary 'Do You Really Want to Know?'," *Shaw TV,* accessed at www.youtube.com/watch?v=j429epbE8MY
194. *The Economist,* "23andme and the FDA," *The Economist,* accessed at http://www.economist.com/news/business/21590941-regulator-brings-genetics-company-halt-and-fda
195. Lynch, Brian. Private communication, November 30, 2013.
196. IBM Corporation. "Pioneering Genetic Privacy," 2011, accessed at http://www-03.ibm.com/ibm/history/ibm100/us/en/icons/geneticprivacy/
197. Botkin, Jeffrey R., *et al.* "Public Attitudes Regarding the Use of Residual Newborn Screening Specimens for Research," *Pediatrics,* 129(2)p. 231-238, February 2012.
198. Bombard, Yvonne, *et al.* "Citizens' Values Regarding Research With Stored Samples From Newborn Screening in Canada," *Pediatrics,*129(2) pp. 239-247, February 2012.
199. Sherwell, Phillip. "Mothers seeking 'Super Donor 401' get a special gift," *The Telegraph,* May 8, 2006, accessed at http://www.telegraph.co.uk/health/health-news/3339259/Mothers-seeking-Super-Donor-401-get-a-special-gift.html
200. Miroz, Jacqueline. "One Sperm Donor, 150 Offspring," *New York Times,* September 5, 2011.
201. *Ibid.*
202. Motluk, Alison. "Tracing Dad Online," *New Scientist,* 188(2524) pp. 6-7, November 5, 2005.
203. Morton, Simon. "Barcelona clubbers get chipped," BBC News, September 24, 2004, accessed at http://news.bbc.co.uk/2/hi/technology/3697940.stm
204. Willard, HF, Ginsburg, GS. *Genomic and Personalized Medicine,* 1st ed. Amsterdam, the Netherlands; Boston, MA: Elsevier/Academic Press, 2009.
205. Davies, Kevin. *The $1000 Genome.* New York: Free Press, 2010.
206. Miller, Tracy. "EXCLUSIVE: Angelina Jolie's doctor says star's double mastectomy will have 'tremendously lasting impact' and save lives," *New York Daily News,* October 14, 2013, accessed at http://www.nydailynews.com/life-style/health/angelina-jolie-doctor-star-double-mastectomy-saving-lives-article-1.1484128
207. American College of Obstetricians and Gynecologists. "ACOG Committee

Opinion: Sex Selection," February 2007, accessed at http://www.acog.org/Resources_And_Publications/Committee_Opinions/Committee_on_Ethics/Sex_Selection
208. Radcliffe, Jerome. "Hacking Medical Devices for Fun and Insulin: Breaking the Human SCADA System," 2011, accessed at http://media.blackhat.com/bh-us-11/Radcliffe/BH_US_11_Radcliffe_Hacking_Medical_Devices_WP.pdf
209. Robertson, Jordan. "Insulin Pumps, Monitors Vulnerable To Hacking," *Huff Post Tech*, August 4, 2011, accessed at http://www.huffingtonpost.com/2011/08/04/insulin-pumps-monitors-vulnerable-to-hacking_n_917987.html
210. IOActive. "IOActive to Reveal Ground-Breaking Research at Black Hat and DEF CON," press release, July 24, 2013, accessed at http://www.ioactive.com/news-events/ioactive_to_reveal_ground_breaking_research_at_black_hat_and_defcon.html
211. City and County of San Francisco. Medical Examiner's Register, Case #2013-0726, accessed at http://www.scribd.com/doc/196531876/Barnaby-Jack-Autopsy-Report
212. AMTV. "INVESTIGATED: Elite Hacker Barnaby Jack Murdered by NSA?" July 29, 2013, accessed at http://www.youtube.com/watch?v=TjHbQJERoso
213. For example, at http://store.neurosky.com/
214. Benchoff, Brian. "Modifying an EEG headset for lucid dreaming," *hackaday.com*, December 21, 2012, accessed at http://hackaday.com/2012/12/20/modifying-an-eeg-headset-for-lucid-dreaming
215. Myndplay. "Myndplay–Train Your Brain through Entertainment," March 17, 2011, accessed at http://www.youtube.com/watch?feature=player_embedded&v=BDCPeaDTTH0
216. Syckodelicus. "Behold Nekomimi Revolution," May 26, 2011, accessed at http://www.youtube.com/watch?v=bUxAwMHgrPQ
217. Ramirez, S., *et al*. "Creating a False Memory in the Hippocampus," *Science* 26 July 2013: 341(6144) pp. 387-391.
218. Defense Advanced Research Projects Agency. "Narrative Networks," n.d., accessed at http://www.darpa.mil/Our_Work/DSO/Programs/Narrative_Networks.aspx
219. Charles River Analytics. "Charles River Analytics to Support DARPA in Researching the Effects of Narratives on Human Behavior," July 30, 2012, accessed at https://www.cra.com/about-us/news.asp?display=detail&id=338
220. Dillow, Clay. "Brain-Scanning Binoculars Harness Soldiers' Unconscious Minds to Locate Threats," Popular Science, July 5, 2012, accessed at http://www.popsci.com/technology/article/2012-07/new-brain-scanning-binoculars-move-mind-melding-battle-tech-closer-deployment
221. Weinberger, Sharon. "Mind control moves into battle," *BBC*, July 5, 2012, accessed at http://www.bbc.com/future/story/20120704-mind-control-moves-into-battle
222. Petro, Lucy S., *et al*. "Decoding face categories in diagnostic subregions of primary visual cortex," February 3, 2013, *European Journal of Neuroscience*, April 2013, 37(7) p. 1130-1139.
223. bengood4000. "Body for sale: How much are your chemical components worth?" March 21, 2011, accessed at http://bgoodscience.wordpress.com/2011/03/21/body-for-sale-how-are-your-chemical-components-worth
224. http://www.oneplusyou.com/bb/cadaver
225. Mareedu, Mouli. "Cadavers on sale at Osmania mortuary," *The New Indian Express*, June 13, 2011, accessed at http://ibnlive.in.com/news/cadavers-on-sale-at-osmania-mortuary/159031-60-121.html
226. Science Care. "Understanding Whole Body Donation," n.d., accessed at http://www.sciencecare.com/body-donation-introduction/
227. Nanyang Technological University. "Plastic realistic: NTU medical students to use plastinated human bodies for anatomy learning," May 23, 2013, accessed at http://media.ntu.edu.sg/NewsReleases/Pages/newsdetail.aspx?news=ad8f3ba0-3fb6-4754-a40e-415ec846b776
228. The Plastination Company, Inc. "The History of Our Body Donation Program," n.d., accessed at http://www.koerperspende.de/en.html
229. Calandrino, Joseph A, *et al*. "Some Consequences of Paper Fingerprinting for Elections," n.d., accessed at http://www.cs.princeton.edu/~jcalandr/papers/paper-evt09.pdf
230. Rueda, Jorge. "Thumbprint readers stir fears in Venezuela vote," *Yahoo! Finance*,

August 6, 2012, accessed at http://finance.yahoo.com/news/thumbprint-readers-stir-fears-venezuela-vote-131434477.html
231. *Vice Japan*. "Fake Funerals in South Korea," July 19, 2013, accessed at http://www.vice.com/en_ca/vice-news/a-good-day-to-die-fake-funerals-in-south-korea
232. Merz, Theo. "LivesOn review: can Twitter make you immortal?" *The Telegraph*, August 16, 2013, accessed at http://www.telegraph.co.uk/technology/mobile-app-reviews/10246708/LivesOn-review-can-Twitter-make-you-immortal.html
233. http://postrapturepetcare.com
234. http://www.themonastery.org/catalog/drofdivinitycertificatedd-p-68.html
235. Kurzweil, Raymond. *The Singularity is Near: When Humans Transcend Biology*. New York: Viking Penguin, 2005.
236. www.highexistence.com/wireheading-a-world-without-negative-emotion
237. Roache, Rebecca. "Enhanced punishment: can technology make life sentences longer?," August 2, 2013, accessed at http://blog.practicalethics.ox.ac.uk/2013/08/enhanced-punishment-can-technology-make-life-sentences-longer/
238. Mears, Bill. "Government can't hold NSA surveillance data longer," *CNN*, March 7, 2014, accessed at http://edition.cnn.com/2014/03/07/politics/nsa-surveillance-extend/
239. Galperin, Eva. "How to Remove Your Google Search History Before Google's New Privacy Policy Takes Effect," blog posting, Feburary 22, 2012, accessed at https://www.eff.org/deeplinks/2012/02/how-remove-your-google-search-history-googles-new-privacy-policy-takes-effect
240. CBC News. "How a Hamilton woman lost then won a $50m lotto ticket," December 5, 2013, accessed at www.cbc.ca/news/canada/hamilton/news/how-a-hamilton-woman-lost-then-won-a-50m-lotto-ticket-1.2449082
241. Cavoukian, Ann. "Guidelines for the Use of Video Surveillance Cameras in Public Places," September 2007, accessed at www.ipc.on.ca/images/Resources/video-e.pdf
242. Accessed at http://quux.org:70/Archives/usenet-a-news/FA.sf-lovers/81.06.23_ucb-vax.1870_fa.sf-lovers.txt
243. Accessed at https://support.google.com/chat/answer/29290?hl=en
244. Black Hat. "Black Hat USA 2013 Keynote - Gen. Alexander," accessed at http://www.youtube.com/watch?v=xvVIZ4OyGnQ
245. Keenan, Thomas P. "Can We Ever Have Technological Security," Canadian Defence and Foreign Affairs Institute and Canadian International Council, October 2013, accessed at http://www.cdfai.org/PDF/CanWeHaveTechnological%20Security.pdf
246. Goldstein, Brett. "When Government Joins the Internet of Things," *New York Times*, September 8, 2013, accessed at http://www.nytimes.com/roomfordebate/2013/09/08/privacy-and-the-internet-of-things/when-government-joins-the-internet-of-things
247. Flagg, Kathryn. "Broken Records? Vermont Takes Its First Steps in the Direction of 'Open Data'," *Seven Days*, October 13, 2013, accessed at http://www.7dvt.com/2013broken-records-vermont-takes-its-first-steps-direction-open-data
248. Saunders, Laura. "Is Your Political Donation Deductible?," *Wall Street Journal*, September 28, 2012, accessed at online.wsj.com/news/articles/SB10000872396390444549204578022201425205738
249. Privacy Rights Clearing House. "Fact Sheet 11: From Cradle to Grave. Government Records and Your Privacy," Revised December 2013, accessed December 6, 2013 at www.privacyrights.org/cradle-grave-government-records-and-your-privacy
250. Calgary Herald. "Tough sledding built up festival founder," *Calgary Herald*, April 27, 2007, accessed at http://www.canada.com/story_print.html?id=23e005c5-2940-46cd-9be9-eb3076d8101f
251. Singer, Natasha. "Mapping, and Sharing, the Consumer Genome," *New York Times*, June 16, 2012, accessed at http://www.nytimes.com/2012/06/17/technology/acxiom-the-quiet-giant-of-consumer-database-marketing.html
252. Acxiom. "PersonicX Cluster, Perspectives," n.d., accessed at http://reference.mapinfo.com/software/anysite_segmentation/english/2_0_1/PersonicX_Binder.pdf
253. The granddaddy of these sites is http://www.wasarrested.com, which has been around since 2001
254. Dewey, Caitlin. "Hunter Moore is in jail, but that just means some other despicable character is 'the most-hated man on the Internet' now," *Washington Post*, January 24, 2014, accessed at http://www.washingtonpost.com/blogs/style-blog/wp/2014/01/24/

hunter-moore-is-in-jail-but-that-just-means-some-other-despicable-character-is-the-most-hated-man-on-the-internet-now/
255. Wlasuk, Alan. "The top 10 hackers of all time," *TechRepublic,* July 26, 2011, accessed at http://www.techrepublic.com/blog/10-things/the-top-10-hackers-of-all-time/
256. Montgomery, James. "Justin Bieber Is Not Going To North Korea, Rep Confirms," *MTV.com,* July 7, 2010, accessed at http://www.mtv.com/news/articles/1643113/justin-bieber-not-going-north-korea-rep-confirms.jhtml
257. Geller, Wendy. "Internet Campaign Helps 'Fat Old Guy' in Quest to Achieve Taylor Swift Dream," *Rolling Stone,* July 18, 2013, accessed at http://www.rollingstone.com/music/news/internet-campaign-helps-fat-old-guy-in-quest-to-achieve-taylor-swift-dream-20130718
258. Topping, David. "Something Awful, Something New," *Torontoist.com,* July 16, 2008 accessed at http://torontoist.com/2008/07/logan_aubes_hockey_night_theme/
259. *The Economist*. "Masters of the cyber-universe," *The Economist,* April 6, 2013, accessed at http://convertpdftoword.net/convertpdftoword.aspx/
260. *Ibid.*
261. Naughton, John. "US fears back-door routes into the net because it's building them too," *The Guardian,* October 13, 2013, accessed at http://www.theguardian.com/technology/2013/oct/13/us-scared-back door-routes-computers-snowden-nsa
262. Simonite, Tom. "A Computer Infection that Can Never Be Cured," *MIT Technology Review,* August 1, 2012, accessed at http://www.technologyreview.com/news/428652/a-computer-infection-that-can-never-be-cured/
263. Karger, Paul A., Schell, Roger R. "Multics Security Evaluation," U.S. Air Force, Electronic Systems Division report ESD-TR-193, Vol. II, June 1974, accessed at csrc.nist.gov/publications/history/karg74.pdf
264. Wetter, Dirk. "Research: Stored XSS Vulnerability @ Amazon," December 18, 2010, accessed at http://drwetter.eu/amazon
265. Love, Dylan. "A Certain String Of Arabic Characters Will Crash Your Mac And iPhone Apps," September 3, 2013, accessed at http://www.businessinsider.com/arabic-characters-ios-bug-2013-9#ixzz2fApGYxUX
266. Collinson, Patrick. "Is this the most wicked scam yet?" *The Guardian*, April 5, 2008, accessed at http://www.theguardian.com/money/2008/apr/05/scamsandfraud
267. Brignull, Harry. "Ryanair trick insurance opt-out trick question (August 2010)," August 4, 2010, accessed at http://darkpatterns.org/ryan-air-trick-insurance-opt-out-trick-question-august-2010/
268. Malware Tips. "Conduit Search–Virus Removal Guide," n.d., accessed at http://malwaretips.com/blogs/remove-conduit-search-virus/
269. Barisani, Andrea, Bianco, Daniele. "Unusual Car Navigation Tricks: Injecting RDS-TMC Traffic Information Signals," Inverse Path, 2007, accessed at http://dev.inversepath.com/rds/cansecwest_2007.pdf
270. Ozer, Nicole. A. "Note to Self: Siri Not Just Working for Me, Working Full-Time for Apple, Too," blog posting, March 12, 2012, accessed at http://www.aclunc.org/issues/technology/blog/note_to_self_siri_not_just_working_for_me,_working_full-time_for_apple,_too.shtml
271. Gay, Alicia. "Apple's Latest iPhone Has No Problem Pointing Users to Viagra, But Comes Up Blank on Birth Control, Abortion," blog posting, November 30, 2011, accessed at www.aclu.org/blog/reproductive-freedom/apples-latest-iphone-has-no-problem-pointing-users-viagra-comes-blank/
272. ProgramYourKeys.com. "BMW 3 Series–Key Remote Programming," n.d., accessed at http://www.programyourkeys.com/BMW_3-Series_Key_Remote_Control_Fob_Programming.html
273. Obias, Rudie. "11 Google Services With Hidden Easter Eggs," Mental Floss, August 20, 2013, accessed at http://mentalfloss.com/article/52191/11-google-services-hidden-easter-eggs
274. Mead, Derek. "How Reddit Got Huge: Tons of Fake Accounts," *Motherboard.tv,* June 21, 2012, accessed at http://motherboard.vice.com/read/how-reddit-got-huge-tons-of-fake-accounts–2
275. Results of this "scam-baiting" can be viewed at http://forum.419eater.com/forum/album.php
276. Defense Distributed. "DD History," n.d., accessed at http://defdist.org/dd-history/
277. Van Zuylen-Wood, Simon. "Philly Becomes First City to Ban 3-D Gun Printing,"

*Philadelphia Magazine*, November 21, 2013, accessed at http://www.phillymag.com/2013/11/21/philly-becomes-first-city-ban-3-d-gun-printing

278. Walters, Ray. "The Pirate Bay declares 3D printed 'physibles' as the next frontier of piracy," *ExtremeTech,* January 24, 2012, accessed at http://www.extremetech.com/electronics/115185-the-pirate-bay-declares-3d-printed-physibles-as-the-next-frontier-of-piracy
279. Wittbrodt, B.T., *et al.* "Life-cycle economic analysis of distributed manufacturing with open-source 3-D printers," *Mechatronics*, June 2013, Elsevier, accessed at www.timelab.org/sites/all/files/studie3Dprinting_0.pdf
280. Daw, David. "Criminals Finding New Uses for 3D Printing," *PC World,* October 10, 2011, accessed at http://www.pcworld.com/article/241605/criminals_find_new_uses_for_3d_printing.html
281. Blatt, Todd. "3D scanning through Glass," blog posting, June 23, 2013, accessed at http://toddblatt.blogspot.ca/2013/06/3d-scanning-through-glass.html
282. Blatt, *Ibid.* "3D scanning through Glass," June 23, 2013, accessed at http://toddblatt.blogspot.com/2013/06/3d-scanning-through-glass.html
283. Weinberg, Michael. "3D Printing, Intellectual Property, and the Fight Over the Next Great Disruptive Technology," excerpt from white paper *at Public Knowledge*, November 2010, accessed at http://p2pfoundation.net/3D_Printing,_Intellectual_Property,_and_the_Fight_Over_the_Next_Great_Disruptive_Technology
284. Keenan, Thomas. "Messing with the Intrinsic Properties of Matter Could Get Very Creepy," decreeping.wordpress.com, February 15, 2013, accessed November 23, 2013 at decreeping.wordpress.com/page/2
285. Lumb, David. "How 3-D-Printing Bones Is Just The Start Of Repairing Your Own Body," *Fast Company,* July 3, 2013, accessed at http://www.fastcolabs.com/3013877/how-3-d-printing-bones-is-just-the-start-of-repairing-your-own-body
286. Innes, Emma, Hodgekiss, Anna. "Cancer left Eric with half his face missing and unable to eat or drink. Now surgeons have made him a new face using a 3D PRINTER," *Mail Online,* April 7, 2013, accessed at http://www.dailymail.co.uk/health/article-2304637/Surgeon-uses-3D-printing-technology-make-cancer-patient-new-face.html
287. 16x9onglobal. "3D Printing: Make anything you want," January 18, 2013, accessed at http://www.youtube.com/watch?v=G0EJmBoLq-g
288. Mims, Christopher. "The audacious plan to end hunger with 3-D printed food," May 21, 2013, accessed at http://qz.com/86685/the-audacious-plan-to-end-hunger-with-3-d-printed-food/
289. New York Toy Collective. "About us," n.d, accessed at http://www.newyorktoycollective.com/about-us
290. IT Business Staff. "11 amazing images show 'The Human Face of Big Data'," *itbusiness.ca*, August 3, 2013, accessed at http://www.itbusiness.ca/slideshows/11-amazing-images-show-the-human-face-of-big-data
291. 3D Printing Industry. "3D Printed Fetuses," *3D Printing Industry,* August 1, 2013, accessed at http://3dprintingindustry.com/2012/08/01/3d-printed-fetuses/
292. LinkedIn. "Our New User Agreement and Privacy Policy," September 12, 2013, accessed at http://blog.linkedin.com/2013/08/19/updates-to-linkedins-terms-of-service/
293. Cluley, Graham. "Shameful and creepy. LinkedIn opens its doors to kids as young as 13 years," blog posting, August 20, 2013, accessed at http://grahamcluley.com/2013/08/linkedin-kids-13-years-old/
294. Rivlin, Jack. "LinkedIn is signing up children–it's official, we've ruined childhood," August 21, 2013, accessed at http://blogs.telegraph.co.uk/technology/jackrivlin/100009930/linkedin-is-signing-up-children-its-official-weve-ruined-childhood/
295. Canadian Press. "Facebook considering opening site to preteens," *CBC News,* October 22, 2013, accessed at http://www.cbc.ca/news/technology/facebook-considers-opening-site-to-preteens-1.2158439
296. Oliver, Gabriel. "The 5 Most Unsettling Disney Theme Park Easter Eggs," July 1997, accessed at www.cracked.com/article_19977_the-5-most-unsettling-disney-theme-park-easter-eggs.html
297. Sato, Munehiko, *et al.* "Touché: Enhancing Touch Interaction on Humans, Screens, Liquids, and Everyday Objects," May 5, 2012, accessed at http://www.disneyresearch.com/wp-content/uploads/touchechi2012.pdf

298. Disney Research. "Ishin-Den-Shin, Transmitting Sound Through Touch," story and video, n.d., accessed at http://www.disneyresearch.com/project/ishin-den-shin/
299. Weiser, M. "The computer for the 21st century." *Scientific American,* (9)September 1991, pp. 94-104.
300. McMahon, Tamsin. "Open secrets online: What's personal and what's corporate?," *Macleans,* March 8, 2014, accessed at http://www.macleans.ca/economy/business/open-secrets-3/
301. Chan, Sewell. "Leona Helmsley's Unusual Last Will," *New York Times City Room* blog posting, August 29, 2007, accessed at http://cityroom.blogs.nytimes.com/2007/08/29/leona-helmsleys-unusual-last-will/
302. Associated Press. "Trouble, the millionaire Maltese, dies," *The Guardian,* June 9, 2011, accessed at http://www.theguardian.com/world/2011/jun/09/trouble-millionaire-maltese-dies
303. Jermyn, Diane. "How to write a 'fur kid' into your will," *Globe and Mail,* July 31, 2012, accessed at http://www.theglobeandmail.com/globe-investor/personal-finance/financial-road-map/how-to-write-a-fur-kid-into-your-will/article4449089/
304. http://www.formalwill.ca/will-products-canada
305. Horn L., *et al.* "The Importance of the Secure Base Effect for Domestic Dogs–Evidence from a Manipulative Problem-Solving Task," *PLoS ONE* 8(5): e65296.
306. Miller, Greg A. *Going … Going … Nuts!: The Story Had to Be Told,* 2004, PublishAmerica, Frederick, MD.
307. http://www.improbable.com/ig/winners/#ig2005
308. Google. "Google Voice Expands to new Markets," blog posting, April 1, 2012, accessed at http://googlevoiceblog.blogspot.ca/2012/04/google-voice-expands-to-new-markets.html
309. Garber, Megan. "Animal Behaviorist: We'll Soon Have Devices That Let Us Talk With Our Pets," *The Atlantic,* June 4, 2013, accessed at http://www.theatlantic.com/technology/archive/2013/06/animal-behaviorist-well-soon-have-devices-that-let-us-talk-with-our-pets/276532/
310. *Ibid.*
311. BBC. "World: Americas: Furby toy or Furby spy?" January 13, 1999, accessed at http://news.bbc.co.uk/2/hi/americas/254094.stm
312. Assange, Julian. *Cypherpunks: Freedom and Future of the Internet.* New York: OR Books, 2012.
313. Hoover, Dwayne. "6 Things You Won't Believe Got Banned by Modern Governments, *cracked.com,* May 12, 2011, accessed at http://www.cracked.com/article_19192_6-things-you-wonE28099t-believe-got-banned-by-modern-governments.html
314. Furby Central. "Furby Cheats," blog posting, n.d., accessed at http://www.virtualpet.com/vp/farm/furby/furby.htm
315. Mimitchi.com. "What does 'Kitty Kitty' mean?," blog posting, n.d, accessed at http://www.mimitchi.com/html/ffaq4.htm
316. BBC. "Disposing of a Furby," blog posting, n.d., accessed at http://www.bbc.co.uk/dna/place-london/plain/A71641
317. Silvanovich, Natalie. "What Will the Tamagotchi Become? and Other Questions of Tamagotchi Life," blog posting, n.d., accessed at http://www.kwartzlab.ca/2013/06/what-will-tamagotchi-become-and-other-questions-tamagotchi-life/
318. Brewster, Signe. "Cyborg insects could map collapsed buildings for first responders," *Gigaom,* October 16, 2013, accessed at http://gigaom.com/2013/10/16/cyborg-insects-could-map-collapsed-buildings-for-first-responders/
319. Backyard Brains. "The Roboroach," n.d., accessed at https://backyardbrains.com/products/roboroach
320. Lockwood, Jeffery. *Six-Legged Soldiers: Using Insects as Weapons of War.* New York: Oxford University Press, 2009.
321. Edwards, Lin. "Cyborg beetles to be U.S. military's latest weapon (w/Video)," *phys.org,* October 15, 2009, accessed at http://phys.org/news174812133.html
322. Brandon, Elicia. "Jules Says Goodbye," video, YouTube, November 20, 2006, accessed at http://www.youtube.com/watch?v=xRR33WDFi_k
323. Annie. "US Army Robots Will Outnumber Human Soldiers in a Decade," blog posting, November 25, 2013, accessed at http://news.filehippo.com/2013/11/us-army-robots-will-outnumber-human-soldiers-in-a-decade/
324. NIMA TV. "Android brothels: robot prostitutes could replace humans by 2050," YouTube, April 23, 2012.

325. Yeoman, Ian, Mars, Michelle. "Robots, Men and Sex Tourism," *Futures*, 44(2012)pp. 365-371, Elsevier, 2011.
326. YouGov. "Omnibus Poll," February 20-21, 2013, accessed at http://big.assets.huffingtonpost.com/toplinesbrobots.pdf
327. Blain, Loz. "[NSFW] Realtouch Interactive: Remote sex is no longer a thing of the future (Part 2)," *gizmag*, May 30, 2013, accessed at http://www.gizmag.com/realtouch-interactive-remote-sex-teledildonics/27441/
328. Efee Link. "App Controlled Interactive Sex Toys for Long Distance Lover Over the Internet," https://www.youtube.com/watch?v=RymJS60xGlo
329. Ikeda, Hayato. "VR Tenga Demo (1): Oculus Rift + Novint Falcon + Tenga," accessed at http://www.youtube.com/watch?v=nLXVinyXjgA
330. Lin, Patrick, *et al*. *Robot Ethics*. Cambridge, MA: MIT Press.
331. Turkle, Sherry. "Sociable Robots," talk at 2013 AAAS meeting, video, at 54:30, accessed at http://www.solvingforpattern.org/2013/02/17/sherry-turkle-aaas-talk-on-sociable-robotics/
332. Erick Schonfeld. "Eric Schmidt Tells Charlie Rose Google Is 'Unlikely' To Buy Twitter And Wants To Turn Phones Into TVs," *TechCrunch*, March 7, 2009, accessed at http://techcrunch.com/2009/03/07/eric-schmidt-tells-charlie-rose-google-is-unlikely-to-buy-twitter-and-wants-to-turn-phones-into-tvs/
333. Champeau, Rachel. "UCLA study finds that searching the Internet increases brain function," *UCLA Newsroom*, October 14, 2008, accessed at http://newsroom.ucla.edu/portal/ucla/ucla-study-finds-that-searching-64348.aspx
334. Small, Gary, Vorgan, Gigi. *iBrain: Surviving the Technological Alteration of the Modern Mind*, New York: Harper Collins, 2008.
335. http://money.cnn.com/2013/04/08/technology/security/shodan/index.html
336. Chazan, M, *et al*. "Microstratigraphic evidence of in situ fire in the Acheulean strata of Wonderwerk Cave, Northern Cape province, South Africa," *PNAS* 2012 109 (20) E1215–E1220.
337. Nissenbaum, Helen. "Privacy as Contextual Integrity," 79 *Washington Law Review*, 119 (2004) pp. 119-157.
338. Madrigal, Alexis C. "The Philosopher Whose Fingerprints Are All Over the FTC's New Approach to Privacy," *The Atlantic*, March 29, 2012, accessed at http://www.theatlantic.com/technology/archive/2012/03/the-philosopher-whose-fingerprints-are-all-over-the-ftcs-new-approach-to-privacy/254365/
339. Bilton, Nick. "Girls Around Me: An App Takes Creepy to a New Level," *New York Times*, March 30, 2012, accessed at http://bits.blogs.nytimes.com/2012/03/30/girls-around-me-ios-app-takes-creepy-to-a-new-level/
340. Jentsch, Ernst Anton. "On the Psychology of the Uncanny," Psychiatrisch-Neurologische Wochenschrift 8.22 (25 Aug. 1906): 195-98 and 8.23 (1 Sept. 1906): 203-05, translated by Roy Sellars, accessed at http://art3idea.psu.edu/locus/Jentsch_uncanny.pdf
341. Colon, Alex. "Foursquare and the future of check-ins," *Gigaom*, March 19, 2014, accessed at http://gigaom.com/2014/03/19/foursquare-and-the-future-of-check-ins/
342. Perez, Sarah. "Social Travel App Jetpac Ditches Facebook, Pivots To Instagram-Based 'City Guides' For At-A-Glance Recommendations," *TechCrunch*, December 5, 2013, accessed at http://techcrunch.com/2013/12/05/social-travel-app-jetpac-ditches-facebook-pivots-to-instagram-based-city-guides-for-at-a-glance-recommendations/
343. Ackerman, Spencer, Ball, James. "Optic Nerve: millions of Yahoo webcam images intercepted by GCHQ," *The Guardian*, February 28, 2014, accessed at http://www.theguardian.com/world/2014/feb/27/gchq-nsa-webcam-images-internet-yahoo
344. Canada Research Chair in Electronic Health Information, accessed at http://www.chairs-chaires.gc.ca/chairholders-titulaires/profile-eng.aspx?profileID=1690
345. Teotonio, Isabel. "Canadian woman denied entry to U.S. because of suicide attempt," *Toronto Star*, January 29, 2011, accessed at http://www.thestar.com/news/gta/2011/01/29/canadian_woman_denied_entry_to_us_because_of_suicide_attempt.html
346. Bradley, Simon, "Whistleblowers and Spying Experts Talk Privacy," Swiss Broadcasting Corporation, October 1, 2013, accessed January 15, 2014 at http://www.swissinfo.ch/eng/politics/Whistleblowers_and_spying_experts_talk_privacy.html
347. Sanger, David E., and Shanker, Thom. "N.S.A. Devises Radio Pathway Into Computers," *New York Times*, January 14, 2014, accessed January 15, 2014 at http://www.nytimes.com/2014/01/15/us/

nsa-effort-pries-open-computers-not-connected-to-internet.html
348. Leafkiller, *et al.* "How to See Who Views Your Facebook Profile the Most," WikiHow, n.d., accessed at http://www.wikihow.com/See-Who-Views-Your-Facebook-Profile-the-Most
349. http://www.unfriendfinderfacebook.com
350. http://www.eyelock.com/index.php/products/myris
351. Monks, Kieron. "The guns that know who is firing them: Can smart tech make firearms safer?," CNN, March 26, 2014, accessed at http://edition.cnn.com/2014/03/26/tech/innovation/smart-guns-know-whos-firing/index.html?hpt=hp_c6
352. Phillips v. Telus Corp., 2002 BCPC 499, in which Lila Dawn Phillips was awarded damages of one dollar because her phone company refused to accept payment of her bill in cash at its retail outlet.
353. Slade, Daryl. "Accused killer searched 'how to dissolve a body'," *Calgary Herald,* December 11, 2013, p. A8.
354. Barbaro, Michael, Zeller, Tom Jr. "A Face Is Exposed for AOL Searcher No. 4417749," *New York Times,* August 9, 2006, accessed at http://www.nytimes.com/2006/08/09/technology/09aol.html?_r=2&
355. http://www.priv.gc.ca/cf-dc/2014/2014_001_0114_e.asp
356. http://classactiondefense.jmbm.com/pineda_class_action_defense_cal.pdf
357. Mabe, Tom. "The Angel of Death cemetery prank call," n.d., accessed at https://www.youtube.com/watch?v=oLnQXmkR3dI
358. Schwartz, John. "Consumers Finding Ways to Zap Telemarketer Calls," New York Times, December 18, 2002, accessed January 15, 2014 at http://www.nytimes.com/2002/12/18/business/consumers-finding-ways-to-zap-telemarketer-calls.html?src=pm
359. Bradford University. "Can Tattoos Cause Harm?," accessed January 15, 2014 at http://www.bradford.ac.uk/about/news/tattoo-risk/?bnr.
360. Tattoo Archive. "Identification," accessed January 15, 2014 at http://www.tattooarchive.com/tattoo_history/identification.html
361. FBI. "Image-Based Matching Technology Offers Identification and Intelligence Prospects," accessed January 15, 2014 at http://www.fbi.gov/about-us/cjis/cjis-link/december-2012/Image-Based%20Matching%20Technology%20Offers%20Identification%20and%20Intelligence%20Prospects
362. Rebuttal evidence in re: *California v. Michael Joe Jackson*, accessed January 15, 2014 at http://www.sbscpublicaccess.org/docs/ctdocs/052505pltmotchandler.pdf
363. Bhardwaj, Julian. "What is your phone saying behind your back?," Naked Security, October 2, 2012, accessed January 15, 2014 at http://nakedsecurity.sophos.com/2012/10/02/what-is-your-phone-saying-behind-your-back/
364. http://www.cockeyed.com/ebay/safeway_card/ebay_copy.html
365. Terdiman, Daniel. "Gaming the Safeway Club Card," *Wired*, July 11, 2003, accessed at http://www.wired.com/techbiz/media/news/2003/07/59589
366. www.directlabs.com, accessed January 15, 2014.
367. As countless presenters have declared in their slide decks: IANAL/IANYL–"I am not a lawyer/I am not your lawyer."
368. New York Law, ypdcrime.com, n.d., accessed at http://ypdcrime.com/penal.law/article190.htm
369. http://www.howtogetitincanada.com/how-to-get-the-us-version-of-itunes-in-canada/
370. Rayman, Graham. "Ken Tarr Launches a Hoax Campaign on an Industry Immune to Shame," *Village Voice,* June 5, 2013, accessed January 15, 2014 at www.villagevoice.com/2013-06-05/news/ken-tarr-reality-tv/full/
371. Applebaum, Yoni. "How the Professor Who Fooled Wikipedia Got Caught by Reddit," *The Atlantic*, May 15, 2012, accessed January 15, 2014 at www.theatlantic.com/technology/archive/2012/05/how-the-professor-who-fooled-wikipedia-got-caught-by-reddit/257134/
372. www.reddit.com/r/AskReddit/comments/sxkig/opinions_please_reddit_do_you_think_my_uncle_joe/c4htt9n
373. Hansell, Saul. "TiVo is Watching When You Don't Watch, and It Tattles," *New York Times*, July 26, 2006, accessed January 15, 2014 at http://www.nytimes.com/2006/07/26/technology/26adco.html?ex=1311566400&en=143cb4893c1c45a9&ei=5090&partner=rssuserland&emc=rss
374. Watson, Julie. "Cecilia Abadie, California Motorist, Cleared In Google Glass Case," January 16,2014, accessed January 18, 2014 at http://www.huffingtonpost.

com/2014/01/16/cecilia-abadie_n_4613257.html
375. Ball, James. "NSA collects millions of text messages daily in 'untargeted' global sweep," *The Guardian*, January 16, 2014, accessed January 16, 2014 at www.theguardian.com/world/2014/jan/16/nsa-collects-millions-text-messages-daily-untargeted-global-sweep
376. www.007voip.com/
377. Schneier, Bruce. "How the NSA Attacks TOR/Firefox Users with QUANTUM AND FOXACID," October 7, 2013, accessed at www.schneier.com/blog/archives/2013/10/how_the_nsa_att.html
378. http://getnarrative.com/
379. Google, Inc. "Introducing our smart contact lens project," Google Official Blog, January 16, 2014, accessed at http://googleblog.blogspot.com/2014/01/introducing-our-smart-contact-lens.html
380. Prey. "Amazing creativity: Use Prey to Track Your Bike," The Prey Blog, n.d., accessed at http://preyproject.com/blog/2012/09/amazing-ingenuity-using-prey-to-track-your-bicycle
381. Huffington Post UK. "'Brainless Thief' Steals Phone Then Uploads Selfie To Whatsapp," Huffington Post UK, January 23, 2014, accessed at http://www.huffingtonpost.co.uk/2013/10/13/brainless-thief-steals-phone-uploads-selfie-whatsapp_n_4094357.html
382. Associated Press, "New York City man sets dating app honey trap to recover iPhone," January 5, 2013, accessed January 15, 2014 at http://www.cbc.ca/news/world/nyc-man-sets-dating-app-honey-trap-to-recover-iphone-1.1304693
383. In re: Google Cookie Placement Consumer Privacy Litigation, 12-md-02358, U.S. District Court, District of Delaware (Wilmington).
384. Sons of Maxwell. "United Breaks Guitars," YouTube, July 6, 2009, accessed at http://www.youtube.com/watch?v=5YGc4zOqozo
385. Pfeiffer, Eric. "Woman gets $3,500 fine and bad credit score for writing negative review of business," Yahoo! News, November 15, 2013, accessed at http://news.yahoo.com/woman-gets–3-500-fine-and-bad-credit-score-for-writing-negative-review-of-business-233833012.html
386. tripper61PA. "Skyauction Scam," Tripadvisor.com, June 28, 2011, accessed at http://www.tripadvisor.ca/ShowTopic-g499445-i9177-k4596925-Skyauction_Scam-Akumal_Yucatan_Peninsula.html
387. http://nyc2600.net/about
388. https://defcon.org/html/defcon-21/dc-21-speakers.html#Keenan
389. http://www.blackhat.com/asia-14/briefings.html#Keenan
390. https://www.defcon.org/html/defcon-21/dc-21-vendors.html